APPLIED RESEARCH IN
FIELD CROP PATHOLOGY
FOR INDIANA, 2020

APPLIED RESEARCH IN FIELD CROP PATHOLOGY FOR INDIANA, 2020

DARCY E. P. TELENKO

PURDUE UNIVERSITY PRESS
WEST LAFAYETTE, INDIANA

Cataloging-in-Publication Data on file at the Library of Congress.

978-1-62671-329-1 (paperback)

978-1-6267-133-0 (epdf)

CONTENTS

ACKNOWLEDGMENTS

This report is a summary of applied field crop pathology research trials conducted in 2020 under the direction of the Purdue Field Crop Pathology program in the Department of Botany and Plant Pathology at Purdue University. The author wishes to thank the Purdue Agronomy Research and Education Center, the Purdue Agricultural Centers, and the many cooperators and contributors who provided the resources needed to support the applied field crop pathology research program in Indiana. Special recognition is extended to Jeffrey Ravellette, Su Shim, and Camila Rocco da Silva for technical skills in managing field trials, data organization and processing, and help preparing this report; Mariama Brown, Tiffanna Ross, Natalia Piñeros, and Audrey Conrad graduate students, who assisted with field trial data collection and analysis; Cayla Haupt, Emily Duncan, and Kaitlin Waibel, undergraduate student interns who assisted with field trial data collection and scouting; Dr. Tom Creswell, Dr. John Bonkowski, and Todd Abrahamson with the Purdue Plant Pest Diagnostic Laboratory for assistance in pathogen surveys and diagnosis; and Dr. Damon Smith and Dr. Daren Mueller for providing peer review. Collectively, the contributions of colleagues, professionals, students, and growers were responsible for a highly successful and productive program to evaluate products and practices for disease management in field crops.

The author would also like to thank the following for their support in 2020: AMVAC, Bayer Crop Science, BASF, Certis USA, Corteva Agriscience, FFAR–ROAR, FMC Agricultural Solution, Gowan, the Indiana Corn Marketing Council, the Indiana Soybean Alliance, the North Central Soybean Research Program, Pioneer, Purdue University, Sipcam Agro, Syngenta, UPD NA Inc., USDA NIFA Hatch Project #1019253, USWBSI—NFO, and Valent.

SUMMARY OF 2020 FIELD CROP DISEASE SEASON

CORN

In 2020, most diseases on corn in Indiana remained relatively low across the state with a few exceptions, as listed below. Gray leaf spot, northern corn leaf blight, northern corn leaf spot, and Diplodia streak were found in pockets. There were also numerous reports of Physoderma brown spot and stalk rot. Tar spot and southern rust were two diseases that were closely monitored this season.

Tar spot: Tar spot of corn was a concern in 2020 following the localized epidemics experienced in 2018 and 2019. In 2020, Indiana continued to have localized epidemics, but they were not as widespread as seen previously. The environmental conditions are key in determining field risk year to year, as leaf wetness plays an important role in tar spot disease development. The second year of tar spot–directed research has been completed here in Indiana. As a cautionary note, it is important to have multiple years of data for verification, but the initial results do serve as a good starting point for making future management decisions.

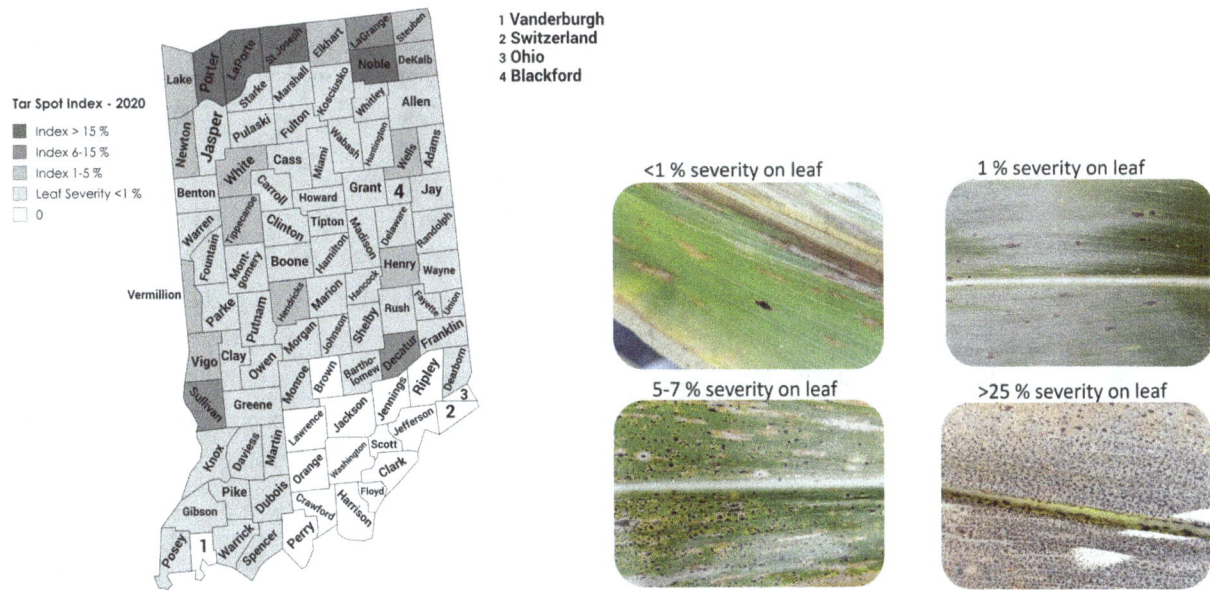

FIGURE 1. 2020 Tar spot index for Indiana. The darker gray the county, the greater the field incidence and severity of tar spot in the fields in which it was found. The range of tar spot severity on leaves was >25%, 5–7%, 1% and <1%. Photo credit: D. Telenko. Map source: https://corn.ipmpipe.org/tarspot/

FIGURE 2. Southern corn rust map of confirmed (gray) counties that had southern corn rust in Indiana in 2020 and a corn leaf with severe southern rust infection. Photo credit: D. Telenko. Map source: https://corn.ipmpipe.org/southerncornrust/.

The field crop pathology team made a large effort at the end of the season to scout for tar spot across the state. Twelve new counties were confirmed with tar spot in 2020, making 78 counties total in Indiana. Out of the 201 fields scouted, 151 were positive for tar spot (75.1%). In addition, incidence and severity were rated (examples of severity are in Figure 1, *right*) and used to generate a tar spot index shown in the map in Figure 1 (*left*)—the darker gray the county, the greater tar spot index observed in 2020. The map demonstrates how corn produced in northern Indiana is at a higher risk for tar spot versus central and southern Indiana, but there are new pockets of disease emerging in central Indiana. The map also parallels the weather conditions and reports during 2020. It is important to document tar spot movement in the state should favorable conditions arise, increasing tar spot disease risk across the remainder of the state.

Southern corn rust: Southern corn rust was first found in Indiana on July 25, 2020, and by the end of the season a total of 59 counties were confirmed (Figure 2, *left*). Southern rust pustules generally tend to occur on the upper surface of the leaf and produce chlorotic symptoms on the underside of the leaf (Figure 2, *right*). These pustules rupture the leaf surface and are orange to tan in color. They are circular to oval in shape.

Common rust was also widespread, and both diseases could be present on a leaf and easily mistaken for each other. It is important to send a sample to the Purdue Plant Pest Diagnostic Lab for confirmation if southern rust is suspected. There is an increased risk for yield impact if southern rust is identified early in the season.

Due to the need to monitor both southern rust and tar spot in Indiana, there will be no charge for Indiana growers to submit southern rust and tar spot samples to the Purdue Plant Pest Diagnostic Lab for diagnostic confirmation. This service is made possible through research supported by the Indiana Corn Marketing Council.

SOYBEAN

Diseases in soybeans remained relatively low throughout the season for much of the state. Our research sites and sentinel plots across the state saw low levels of frogeye leaf spot, Cercospora leaf blight, downy mildew, and Septoria brown spot. There were a few spots of sudden death syndrome and white mold as well. In general, it was a quiet year for diseases in soybean.

WHEAT

Fusarium head blight (FHB), or scab, is one of the most impactful diseases of wheat and the most challenging to prevent. In addition, FHB infection can cause the production of a mycotoxin called deoxynivalenol (DON, or vomitoxin). The conditions in 2020 were moderately conducive to FHB development. Our research sites in both West Lafayette and Vincennes had moderate levels of FHB develop in our nontreated susceptible cultivar checks, and initial DON testing was less than 1 ppm. Fusarium head blight management requires an integrated approach. This includes selection of cultivars with moderate resistance and timely fungicide application at flowering. Other diseases observed in our wheat trials in 2020 included leaf rust, Septoria leaf and glume blotch, and stripe rust.

AGRONOMY CENTER FOR RESEARCH AND EDUCATION (ACRE)

COMPARISON OF FUNGICIDES APPLIED AT TASSEL/SILK (VT/R1) OR MILK (R3) GROWTH STAGES FOR FOLIAR DISEASES IN CORN IN CENTRAL INDIANA, 2020 (COR20-01.ACRE)

S. Shim, J. D. Ravellette, and D. E. P. Telenko, Department of Botany and Plant Pathology, Purdue University West Lafayette, Indiana 47907-2054

CORN (ZEA MAYS P9998AM)

Gray leaf spot, *Cercospora zeae-maydis*

A research trial was established at the Purdue Agronomy Center for Research and Education (ACRE) in Tippecanoe County, Indiana. The trial was a randomized complete block design with four replications. Plots were 10 feet wide and 30 feet long and consisted of four rows, and the two center rows were used for evaluation. The previous crop was corn. Standard practices for nonirrigated grain corn production in Indiana were followed. Corn hybrid P9998AM was planted in 30-inch row spacing at a rate of 34,000 seeds/acre on May 25. All fungicide applications were applied at 15 gal/acre at 40 psi using a Lee self-propelled sprayer equipped with a 10-foot boom, fitted with six TJ-VS 8002 nozzles spaced 20 inches apart, at 3.6 mph. In-furrow treatments were applied while planting on May 24. Fungicides were applied on July 25 at tassel/silk (VT/R1) growth and on August 18 at milk (R3) growth stage. Disease ratings were assessed on August 25 at beginning dent (R5) and September 9 at late maturity (R6) growth stages. Disease severity was visually assessed as a percentage (0–100%) of symptomatic leaf area on ear leaf. Five plants were assessed per plot and averaged before analysis. The two center rows of each plot were harvested on October 6, and yields were adjusted to 15.5% moisture. Data were subjected to mixed model analysis of variance (SAS 9.4, 2019), and means were compared using Fisher's Least Significant Difference test (LSD; $\alpha=0.05$).

In 2020, gray leaf spot, northern corn leaf blight, and Physoderma brown spot and stalk rot were the most prominent diseases in the trial and reached low severity. All fungicide programs significantly reduced gray leaf spot severity on the ear leaf compared to the nontreated controls on August 25 and September 9 except

Veltyma at R3 on August 25 (Table 1). All fungicides increased the percentage of green canopy over the non-treated controls. Lucento at VT/R1 and R3, Veltyma at VT/R1, and Xyway in-furrow resulted in greener plots than Veltyma at R3, Delaro at R3, and Quilt Xcel at R3, but these were not significantly different from any of the other treatments. There was no significant difference between treatments for harvest moisture, test weight, and corn yield.

TABLE 1. *Effect of Fungicide on Foliar Disease Severity and Corn Yield*

TREATMENT, RATE/ACRE, AND TIMING[z]	GLS[y]% AUGUST 25	GLS[y]% SEPTEMBER 9	CANOPY GREEN[x]% SEPTEMBER 9	HARVEST MOISTURE %	TEST WEIGHT LB/BU	YIELD[w] BU/ACRE
Nontreated control	1.8 ab	7.9 a	73.8 c	21.5	55.0	198.6
Lucento 4.17 SC 5.0 fl oz at VT/R1	0.2 fg	0.3 d	95.0 a	20.4	54.6	202.4
Trivapro 2.21 SE 13.7 fl oz at VT/R1	0.3 ef	0.6 d	87.5 abc	21.3	54.4	195.7
Miravis Neo 2.5 SE 13.7 fl oz at VT/R1	0.4 deg	0.3 d	93.8 ab	21.6	54.2	217.8
Veltyma 3.34 S 7.0 fl oz at VT/R1	0.1 fg	0.1 d	95.0 a	21.2	54.3	193.3
Delaro 325 SC 8.0 fl oz at VT/R1	0.2 fg	0.6 d	93.8 ab	21.7	54.5	216.5
Quilt Xcel 2.2 SE 10.5 fl oz at VT/R1	0.5 def	1.1 d	91.3 abc	20.9	54.4	196.5
Headline AMP 1.68 SC 10.0 fl oz at VT/R1	0.2 fg	0.7 d	90.0 abc	21.8	54.4	204.0
Revytek 3.33 LC 8.0 fl oz at VT/R1	0.7 c-g	0.2 d	93.8 ab	22.3	53.3	217.7
Xyway LFR 15.2 fl oz in-furrow	0.2 fg	0.7 d	95.0 a	21.1	57.7	193.5
Xyway LFR 8.35 fl oz in-furrow fb Lucento 4.17 SC 5.0 fl oz at VT/R1	0.1 g	0.2 d	90.0 abc	21.9	54.1	208.4
Lucento 4.17 SC 5.0 fl oz at R3	0.6 d-g	0.9 d	95.0 a	20.9	54.7	205.7
Trivapro 2.21 SE 13.7 fl oz at R3	1.0 b-g	1.3 d	87.5 abc	21.4	54.8	209.5
Miravis Neo 2.5 SE 13.7 fl oz at R3	1.0 a-d	0.7 d	87.5 abc	20.8	54.3	203.2
Veltyma 3.34 S 7.0 fl oz at R3	2.2 a	1.7 cd	83.8 c	21.5	54.9	204.5
Delaro 325 SC 8.0 fl oz at R3	1.2 a-e	3.7 bc	83.8 c	21.2	54.5	203.9
Quilt Xcel 2.2 SE 10.5 fl oz at R3	1.1 b-f	1.6 d	86.3 bc	21.1	54.1	211.3
Headline AMP 1.68 SC 10.0 fl oz at R3	0.8 c-g	1.1 d	87.5 abc	21.2	54.6	201.0
Revytek 3.33 LC 8.0 fl oz at R3	1.6 abc	1.3 d	83.8 c	20.9	54.5	215.9
Nontreated control	0.7 c-g	5.3 b	75.0 d	21.4	54.1	198.6
P-value	*0.0006*	*<.0001*	*<.0001*	*0.0539*	*0.9097*	*0.5520*

[z] Fungicide treatments were applied on July 25 at tassel/silk (VT/R1) and August 18 at milk (R3) growth stage. All treatments contained a nonionic surfactant (Preference) at a rate of 0.25% v/v. fb = followed by.

[y] Disease severity was visually assessed as a percentage (0–100%) of symptomatic leaf area on ear leaf; five plants were assessed per plot and averaged before analysis. GLS = gray leaf spot.

[x] Canopy green was visually assessed as a percentage (0–100%) of crop canopy on September 9.

[w] Yields were adjusted to 15.5% moisture at harvest on October 6.

[v] Means followed by the same letter are not significantly different based on Fisher's Least Significant Difference test (LSD; α=0.05).

UNIFORM FUNGICIDE TRIAL FOR TAR SPOT AND OTHER FOLIAR DISEASES IN CORN IN CENTRAL INDIANA, 2020 (COR20-02.ACRE)

T. J. Ross, J. D. Ravellette, S. Shim, and D. E. P. Telenko, Department of Botany and Plant Pathology, Purdue University West Lafayette, Indiana 47907-2054

CORN (*ZEA MAYS* W2585SSRIB)

Gray leaf spot, *Cercospora zeae-maydis*
Tar spot, *Phyllachora maydis*

A trial was established at the Purdue Agronomy Center for Research and Education (ACRE) in Tippecanoe County, Indiana. The experiment was a randomized complete block design with four replications. Plots were 10 feet wide and 30 feet long and consisted of four rows, and the two center rows were used for evaluation. The previous crop was corn. Standard practices for nonirrigated grain corn production in Indiana were followed. Corn hybrid W2585SSRIB was planted in 30-inch row spacing at a rate of 34,000 seeds/acre on May 25 using a John Deere 1700 six-row planter. All fungicide applications were applied at 15 gal/acre at 40 psi using a Lee self-propelled sprayer equipped with a 10-foot boom, fitted with six TJ-VS 8002 nozzles spaced 20 inches apart, at 3.6 mph. Fungicides were applied on July 25 at tassel/silk (VT/R1) growth stage. Disease ratings were assessed on August 25 and September 23 at the early dough (R4) and dent (R5) growth stages, respectively. Disease severity was visually assessed as a percentage (0–100%) of symptomatic leaf area on in the lower and upper canopy. The two center rows of each plot were harvested on October 18, and yields were adjusted to 15.5% moisture. Data were subjected to mixed model analysis of variance (SAS 9.4, 2019), and means were compared using Fisher's Least Significant Difference test (LSD; α=0.05).

Gray leaf spot (GLS) and tar spot were the most prominent diseases in the trial and reached low severity. All fungicide treatments reduced the severity of GLS and tar spot and increased the percentage of green canopy over the nontreated control (Table 2). Headline Amp had the highest percent of green canopy on September 23 but was only significantly different from Tilt. There was no significant difference between treatments for harvest moisture, test weight, and yield.

TABLE 2. *Effect of Fungicide on Foliar Disease Severity and Corn Yield*

TREATMENT, RATE/ACRE[z]	GLS[y] % AUGUST 25	TAR SPOT[y] % SEPTEMBER 23	CANOPY GREEN[x] SEPTEMBER 23	HARVEST MOISTURE %	TEST WEIGHT LB/BU	YIELD[w] BU/ACRE
Nontreated control	15.0 a	0.10 a	48.8 c	18.7	55.6	227.9
Revytek 3.33 LC 8.0 fl oz	5.0 b	0.03 b	65.0 ab	18.8	55.4	220.0
Veltyma 3.34 SC 7.0 fl oz	4.5 b	0.01 b	62.5 ab	18.8	56.0	227.0
Headline 2.08 SC 6.0 fl oz	5.8 b	0.03 b	63.8 ab	18.9	55.9	230.0
Headline AMP 1.68 SE 10.0 fl oz	5.5 b	0.03 b	68.8 a	18.9	56.4	222.2
Aproach Prima 2.34 SC 6.8 fl oz	5.5 b	0.01 b	65.0 ab	19.1	56.3	226.1
Miravis Neo 2.5 SE 13.7 fl oz	6.3 b	0.01 b	60.0 abc	18.3	56.1	222.6
Delaro 325 SC 8.0 fl oz	7.5 b	0.01 b	53.8 bc	18.8	56.5	219.7
Lucento 4.1 SC 5.0 fl oz	5.5 b	0.03 b	62.5 ab	18.5	56.5	222.9
Tilt 3.6 EC 4.0 fl oz	5.5 b	0.03 b	55.0 bc	19.1	56.2	223.1
P-value[v]	<.0001	0.0045	0.0346	0.8332	0.2581	0.6274

[z] Fungicide treatments were applied on July 25 at tassel/silk (VT/R1) growth stage, and all treatments contained a nonionic surfactant (Preference) at a rate of 0.25% v/v.

[y] Disease severity was visually assessed as a percentage (0–100%) of symptomatic leaf area on August 25. GLS = gray leaf spot.

[x] Canopy green was visually assessed as a percentage (0–100%) of crop canopy green on September 23.

[w] Yields were adjusted to 15.5% moisture at harvest on October 18.

[v] Means followed by the same letter are not significantly different based on Fisher's Least Significant Difference test (LSD; α=0.05).

FUNGICIDE TIMING AND MODEL VALIDATION FOR TAR SPOT OF CORN IN CENTRAL INDIANA, 2020 (COR20-04.ACRE)

C. Rocca Da Silva, J. D. Ravellette, S. Shim, and D. E. P. Telenko, Department of Botany and Plant Pathology, Purdue University West Lafayette, Indiana 47907-2054

CORN (*ZEA MAYS* W2585SSRIB)

Gray leaf spot, *Cercospora zeae-maydis*
Tar spot, *Phyllachora maydis*

A trial was established at the Purdue Agronomy Center for Research and Education (ACRE) in Tippecanoe County, Indiana. The experiment was a randomized complete block design with four replications. Plots were 10 feet wide and 30 feet long and consisted of four rows, and the two center rows were used for evaluation. The previous crop was corn. Standard practices for nonirrigated grain corn production in Indiana were followed. Corn hybrid W2585SSRIB was planted in 30-inch row spacing at a rate of 34,000 seeds/acre on May 25 using a John Deere 1700 six-row planter. All fungicide applications were applied at 15 gal/acre at 40 psi using a Lee self-propelled sprayer equipped with a 10-foot boom, fitted with six TJ-VS 8002 nozzles spaced 20 inches apart, at 3.6 mph. Fungicides were applied on July 1 at V8, on July 13 at V10, on July 25 on tassel/silk (VT/R1), on August 9 at blister (R2), on August 18 at milk (R3), on August 25 at dough (R4), and on September 9 at dent (R5) growth stages. No applications were made based on the tar spot risk model (https://connect. doit.wisc.edu/cpn-risk-tool/). Gray leaf spot (GLS) was rated by visually assessing as a percentage (0–100%) severity of disease on lower canopy on August 25 at R3 growth stage. Tar spot was assessed on September 15 at the R5 growth stage. Tar spot was rated by visually by assessing the percentage of stromata per leaf on five plants in each plot at the ear leaf. Values for each plot were averaged before analysis. The two center rows of each plot were harvested on October 18, and yields were adjusted to 15.5% moisture. Data were subjected to mixed model analysis of variance (SAS 9.4, 2019), and means were compared using Fisher's Least Significant Difference test (LSD; α=0.05).

Gray leaf spot (GLS) and tar spot were the most prominent diseases in the trial and reached low severity. All fungicide treatments reduced the severity of GLS on August 25 in the lower canopy and tar spot on September 15 on ear leaf over the nontreated control (Table 3). There was no significant difference between treatments for canopy greenness, harvest moisture, test weight, and yield.

TABLE 3. *Effect of Fungicide on Foliar Disease Severity and Yield*

TREATMENTS, RATE/A AND TIMING[z]	GLS % AUGUST 25	TAR SPOT % SEPTEMBER 15	CANOPY GREEN % SEPTEMBER 23	HARVEST MOISTURE %	TEST WEIGHT LB/BU	YIELD[v] BU/ACRE
Nontreated control	9.3 a	0.3 a	58.8	19.3	55.9	215.2
Trivapro 2.21 SE 13.7 fl oz at V8	1.5 e	0.3 ab	46.3	19.5	55.9	212.0
Trivapro 2.21 SE 13.7 fl oz at V10	2.8 de	0.3 abc	60.0	20.1	56.0	209.5
Trivapro 2.21 SE 13.7 fl oz at VT/R1	4.3 cd	0.2 bcd	63.8	19.6	56.0	208.2
Trivapro 2.21 SE 13.7 fl oz at R2	5.5 bc	0.1 de	66.3	19.8	55.9	206.4
Trivapro 2.21 SE 13.7 fl oz at R3	5.0 c	0.0 e	61.3	19.7	55.2	210.6
Trivapro 2.21 SE 13.7 fl oz at R4	4.3 cd	0.0 e	58.8	19.8	56.0	209.7
Trivapro 2.21 SE 13.7 fl oz at R5	4.5 cd	0.1 cde	60.0	19.8	56.0	203.3
Trivapro 2.21 SE 13.7 fl oz at V8 fb VT	1.0 e	0.4 a	60.0	19.3	56.0	203.4
Trivapro 2.21 SE 13.7 fl oz at tar spot model (no application)	7.0 b	0.2 abc	51.3	18.8	56.2	207.0
P-value[u]	<.0001	0.0002	0.3830	0.5399	0.3535	0.9046

[z] Fungicide treatments were applied on July 1 at vegetative 8-leaf (V8), on July 13 at vegetative 10-leaf (V10), on July 25 on the tassel/silk (VT/R1), on August 9 at the R2 (blister), on August 18 at R3 (milk), on August 25 at dough (R4), and on September 9 at dent (R5) growth stages. No tar spot model application was used, as the model never crossed the threshold. All treatments contained a nonionic surfactant (Preference) at a rate of 0.25% v/v. fb = followed by.

[y] Disease severity was visually assessed as a percentage (0–100%) of lower canopy on August 25. GLS = gray leaf spot.

[x] Tar spot stromata was visually assessed as a percentage (0–100%) of leaf area on five plants in each plot at the ear leaf.

[w] Canopy greenness was visually assessed as a percentage (0–100%) of crop canopy green on September 23.

[v] Yields were adjusted to 15.5% moisture at harvest on October 18.

[u] Means followed by the same letter are not significantly different based on Fisher's Least Significant Difference test (LSD; α=0.05).

EVALUATION OF FUNGICIDES FOR FOLIAR DISEASE IN CORN IN CENTRAL INDIANA, 2020 (COR20-19.ACRE)

S. Shim, J. D. Ravellette, and D. E. P. Telenko, Department of Botany and Plant Pathology, Purdue University West Lafayette, Indiana 47907-2054

CORN (*ZEA MAYS* P9998AM)

Gray leaf spot, *Cercospora zeae-maydis*

A trial was established at the Purdue Agronomy Center for Research and Education (ACRE) in Tippecanoe County, Indiana. The trial was a randomized complete block design with four replications. Plots were 10 feet wide and 30 feet long and consisted of four rows, and the two center rows were used for evaluation. The previous crop was corn. Standard practices for nonirrigated grain corn production in Indiana were followed. Corn hybrid P9998AM was planted in 30-inch row spacing at a rate of 34,000 seeds/acre on May 25. All fungicide applications were applied at 15 gal/acre at 40 psi using a Lee self-propelled sprayer equipped with a 10-foot boom, fitted with six TJ-VS 8002 nozzles spaced 20 inches apart, at 3.6 mph. Fungicides were applied on June 24 at V5/V6, July 17 at V12, July 25 at silk (R1), and August 9 at blister (R2) growth stages. Disease ratings were assessed on August 25 at dent (R5) and September 9 at maturity (R6) growth stages. Disease severity was visually assessed as a percentage (0–10 0%) of symptomatic leaf area on ear leaf; five plants were assessed per plot and averaged before analysis. The two center rows of each plot were harvested on October 6, and yields were adjusted to 15.5% moisture. Data were subjected to mixed model analysis of variance (SAS 9.4, 2019), and means were compared using Fisher's Least Significant Difference test (LSD; α=0.05).

In 2020, gray leaf spot (GLS) was the most prominent diseases in the trial and reached low severity. All fungicides significantly reduced GLS severity over the nontreated control by September 9 except Delaro Complete 458 SC applied at V5, which was not different from the nontreated control on August 25 and had significantly more disease than all other treatments on September 9 (Table 4). Harvest moisture was significantly higher under Trivapro, Miravis Neo, and Veltyma treatments. There was no significant difference between treatments for test weight and corn yield.

TABLE 4. *Effect of Fungicide on Foliar Disease and Corn Yield*

TREATMENT, RATE/ACRE, AND TIMING[z]	GLS[y] % AUGUST 25	GLS[y] % SEPTEMBER 9	HARVEST MOISTURE %	TEST WEIGHT LB/BU	YIELD[x] BU/ACRE
Nontreated control	1.1 a	2.5 a	21.1 b	54.0	195.7
Miravis Neo 2.5 SE 13.7 fl oz at V12	0.0 c	0.1 c	22.3 ab	54.0	206.2
Trivapro 2.21 SE 13.7 fl oz at V12	0.2 bc	0.4 c	22.6 a	53.4	200.9
Miravis Neo 2.4 SE 13.7 fl oz at R1	0.1 bc	0.1 c	22.7 a	53.8	211.3
Trivapro 2.21 SE 13.7 fl oz at R1	0.1 bc	0.3 c	22.2 ab	53.5	205.7
Miravis Neo 2.4 SE 13.7 fl oz at R2	0.5 b	0.4 c	22.2 ab	53.6	205.2
Delaro Complete 458 SC 4.0 fl oz at V5	1.1 a	1.1 b	21.9 ab	54.6	200.1
Delaro Complete 458 SC 8.0 fl oz at R1	0.2 bc	0.3 c	22.1 ab	53.5	195.2
Veltyma 3.34 S 7.0 fl oz at R1	0.1 bc	0.1 c	23.1 a	53.3	194.5
P-value	<.0001	<.0001	0.0197	0.3910	0.7295

[z] Fungicide treatments were applied on June 24 at V5/V6, July 17 at V12, July 25 at silk (R1), and August 9 at blister (R2) growth stages. All treatments applied at R1 and R2 contained a nonionic surfactant (Preference) at a rate of 0.25% v/v.

[y] Disease severity was visually assessed as a percentage (0–100%) of symptomatic leaf area on ear leaf; five plants were assessed per plot and averaged before analysis. GLS = gray leaf spot.

[x] Yields were adjusted to 15.5% moisture at harvest on October 6.

[w] Means followed by the same letter are not significantly different based on Fisher's Least Significant Difference test (LSD; α=0.05).

EVALUATION OF FUNGICIDES FOR FOLIAR DISEASES ON SOYBEAN IN CENTRAL INDIANA, 2020 (SOY20-01.ACRE)

N. Piñeros-Guerrero, J. D. Ravellette, and D. E. P. Telenko, Department of Botany and Plant Pathology, Purdue University West Lafayette, Indiana 47907-2054

SOYBEAN (*GLYCINE MAX* P34A79X)

Cercospora leaf blight, *Cercospora kikuchii/ C. flagellaris*
Septoria brown spot, *Septoria glycines*

A trial was established at the Purdue Agronomy Center for Research and Education (ACRE) in Tippecanoe County. The experiment was a randomized complete block design with four replications. Plots were 10 feet wide and 30 feet long and consisted of four rows, and the two center rows were used for evaluation. The previous crop was corn. Standard practices for soybean production in Indiana were followed. Soybean cultivar P34A79X was planted in 30-inch row spacing at a rate of 140,000 seeds/acre on June 2. All fungicide applications were applied at 15 gal/acre at 40 psi using a Lee self-propelled sprayer equipped with a 10-foot boom, fitted with six TJ-VS 8002 nozzles spaced 20 inches apart, at 3.6 mph. Fungicides were applied on July 29 at the beginning pod (R3) growth stage. Disease ratings were assessed on August 25 at beginning seed (R5) and September 9 at full seed (R6) growth stages. Frogeye leaf spot (FLS), Cercospora leaf blight (CLB), and Septoria brown spot (SBS) were rated for disease severity by visually assessing the percentage of symptomatic leaf area in the upper and lower canopies, respectively. The two center rows were harvested on October 16, and yields were adjusted to 13% moisture. Data were subjected to mixed model analysis of variance (SAS 9.4, 2019), and means were compared using Fisher's Least Significant Difference test (LSD; α=0.05).

In 2020, weather conditions were unfavorable for FLS. SBS and CLB were the most prominent diseases in the trial and reached low severity. All fungicide treatments significantly reduced SBS severity over the nontreated control on both August 25 and September 9. All fungicide treatments significantly reduced CLB severity over the nontreated control on September 9 except for Quadris Top SBX, Lucento, and Trivapro. (Table 5). No significant treatment differences were detected for soybean test weight and yield.

TABLE 5. *Effect of Fungicide on Foliar Disease Severity and Soybean Yield*

TREATMENT, RATE/ACRE[z]	SBS[y] % AUGUST 25	SBS[y] % SEPTEMBER 9	CLB[y] % SEPTEMBER 9	HARVEST MOISTURE %	TEST WEIGHT LB/BU	YIELD[x] BU/ACRE
Nontreated control	3.1 a	8.8 a	0.4 ab	10.3 bc	56.3	62.9
Preemptor 3.22 SC 5.0 fl oz	1.1 b	0.9 c	0.0 c	10.5 ab	56.1	60.7
Topguard EQ 4.29 SC 5.0 fl oz	0.3 b	1.6 bc	0.0 c	10.3 ab	55.9	60.9
Quadris Top SBX 3.76 SC 7.0 fl oz	0.6 b	0.6 c	0.2 bc	10.1 c	56.3	66.6
Lucento 4.17 SC 5.0 fl oz	0.3 b	0.9 c	0.3 bc	10.1 c	56.4	64.9
Miravis Top 1.67 SC 13.7 fl oz	0.1 b	0.2 c	0.0 c	10.1 c	56.3	66.5
Priaxor Xemium SC 4.0 fl oz	0.6 b	0.8 c	0.0 c	10.3 abc	56.2	64.6
Trivapro 2.21 SE 13.0 fl oz	1.0 b	0.8 c	0.5 a	10.3 bc	56.1	61.0
Delaro 325 SC 8.0 fl oz	0.4 b	0.9 c	0.0 c	10.7 a	56.4	62.3
Headline AMP 1.68 SC 10.0 fl oz	1.0 b	3.0 b	0.0 c	10.4 abc	56.2	62.3
Veltyma 3.34 S 7.0 fl oz	0.3 b	0.8 c	0.1 c	10.2 bc	56.3	64.9
Revytek 3.33 LC 8.0 fl oz	0.2 b	0.4 c	0.0 c	10.4 abc	56.2	61.2
P-value[w]	0.0002	<.0001	0.0044	0.0458	0.8710	0.4602

[z] Fungicide treatments were applied on July 29 at the beginning pod (R3) growth stage, and all treatments contained a nonionic surfactant (Preference) at a rate of 0.25% v/v.

[y] Disease severity was visually assessed as a percentage (0–100%) of symptomatic leaf area on August 25 and September 9. SBS = Septoria brown spot, CLB = Cercospora leaf blight.

[x] Yields were adjusted to 13% moisture at harvest on October 16.

[w] Means followed by the same letter are not significantly different based on Fisher's Least Significant Difference test (LSD; α=0.05).

EVALUATION OF FUNGICIDES FOR FOLIAR DISEASES ON SOYBEAN IN CENTRAL INDIANA, 2020 (SOY20-13.ACRE)

D. E. P. Telenko, J. D. Ravellette, and S. Shim, Department of Botany and Plant Pathology, Purdue University West Lafayette, Indiana 47907-2054

SOYBEAN (*GLYCINE MAX* P34A79X)

Frogeye leaf spot, *Cercospora sojina*
Cercospora leaf blight, *Cercospora kikuchii/C. flagellaris*
Septoria brown spot, *Septoria glycines*

A trial was established at the Purdue Agronomy Center for Research and Education (ACRE) in Tippecanoe County. The experiment was a randomized complete block design with four replications. Plots were 10 feet wide and 30 feet long and consisted of four rows, and the two center rows were used for evaluation. The previous crop was corn. Standard practices for soybean production in Indiana were followed. Soybean cultivar P35T75X was planted in 30-inch row spacing at a rate of 140,000 seeds/acre on June 2. All fungicide applications were applied at 15 gal/acre at 40 psi using a Lee self-propelled sprayer equipped with a 10-foot boom, fitted with six TJ-VS 8002 nozzles spaced 20 inches apart, at 3.6 mph. Fungicides were applied on July 29 at the beginning pod (R3) growth stage. Disease ratings were assessed on September 9 at the full seed (R6) growth stage. Frogeye leaf spot (FLS), Cercospora leaf blight (CLB), and Septoria brown spot (SBS) were rated for disease severity by visually assessing the percentage of symptomatic leaf area in the upper and lower canopies. The two center rows were harvested on October 14, and yields were adjusted to 13% moisture. Data were subjected to mixed model analysis of variance (SAS 9.4, 2019), and means were separated using Fisher's Least Significant Difference (LSD; α=0.05).

In 2020, weather conditions were not favorable for soybean disease. FLS, SBS, and CLB were the most prominent diseases in the trial and reached low severity. All fungicides reduced FLS, SBS and CLB on September 9 over the nontreated control (Table 6). Miravis top and Revytek had the lowest amount of SBS but were not significantly different from Lucento, Lucento plus Quadris, and Delaro. No significant treatment effects were detected for soybean harvest moisture, test weight, and yield.

TABLE 6. *Effect of Fungicide on Foliar Disease Severity*

TREATMENT, RATE/ACRE[z]	FLS[y] % SEPTEMBER 9	SBS[y] % SEPTEMBER 9	CLB[y] % SEPTEMBER 9	HARVEST MOISTURE %	TEST WEIGHT LB/BU	YIELD[x] BU/A
Nontreated control	0.1 a	7.5 a	0.1 a	11.4	56.5	54.5
Topguard EQ 5.0 fl oz	0.0 b	2.5 b	0.0 b	11.5	56.4	58.0
Lucento 4.17 SC 5.0 fl oz	0.0 b	1.9 bc	0.0 b	11.3	56.7	57.9
Lucento 4.17 SC 5.0 fl oz + Quadris 2.08 SC 6.0 fl oz	0.0 b	0.9 bc	0.0 b	11.4	56.5	55.5
Miravis Top 1.67 SC 13.7 fl oz	0.0 b	0.1 c	0.0 b	11.4	56.1	57.9
Revytek 3.33 LC 8.0 fl oz	0.0 b	0.5 c	0.0 b	11.5	56.7	56.9
Delaro 325 SC 8.0 fl oz	0.0 b	0.8 bc	0.0 b	11.5	56.5	54.5
P-value	<.0001	<.0001	0.0327	0.5540	0.7546	0.6860

[z] Fungicide treatments were applied on July 29 at the R3 growth stage, and all treatments contained a nonionic surfactant (Preference) at a rate of 0.25%.

[y] Foliar disease incidence was rated on a scale of 0–100% of plants within a plot with disease symptoms on September 9. FLS = frogeye leaf spot, SBS = Septoria brown spot, CLB = Cercospora leaf blight.

[x] Yields were adjusted to 13% moisture at harvest on October 14.

[w] Means followed by the same letter are not significantly different based on Fisher's Least Significant Difference (LSD; α=0.05).

COMPARE THE EFFICACY OF SEED TREATMENTS IN SOYBEAN IN CENTRAL INDIANA, 2020 (SOY20-17.ACRE)

S. Shim, J. D. Ravellette, and D. E. P. Telenko, Department of Botany and Plant Pathology, Purdue University West Lafayette, Indiana 47907-2054

SOYBEAN (*GLYCINE MAX* P25A27X AND P24T76E)

Sudden death syndrome, *Fusarium virguliforme*
Soybean cyst nematode, *Heterodera glycines*

A trial was established at the Purdue Agronomy Center for Research and Education (ACRE) in Tippecanoe County. The experiment was a randomized complete block design with four replications. Plots were 10 feet wide and 30 feet long and consisted of four rows, and the two center rows were used for evaluation. The previous crop was corn. Standard practices for soybean production in Indiana were followed. Soybean cultivar P25A27X (resistant) and P24T76E (susceptible) were planted in 30-inch row spacing at a rate of 8 seeds/foot on June 5. Seed treatments were applied on seeds before planting. Ten roots per plot were sampled from border rows at full pod (R4) on August 24, gently washed, and rated for root rot severity on a scale of 0–100%. Disease ratings were assessed on August 25 at beginning pod/full pod (R3/R4) growth stage. Sudden death syndrome (SDS) in each plot was rated for disease incidence (DI) and disease severity (DS). DI refers to the percentage of plants with disease symptoms, and DS was rated using a 1–9 scale where 1 refers to low disease pressure and 9 refers to premature death of the plant. The SDS index was then calculated using the equation DX = (DI x DS)/9. The two center rows of each plot were harvested on October 14, and yields were adjusted to 13% moisture. Data were subjected to mixed model analysis of variance (SAS 9.4, 2019), and means were compared using Fisher's Least Significant Difference test (LSD; α=0.05).

In 2020, very little disease developed in plots. SDS was the most prominent disease in the trial but only reached low severity. Soybean cyst nematode egg count in spring soil samples ranged from 300 to 700 eggs/100 cc soil, a low to moderate range. There was no significant difference between seed treatments for SDS incidence and severity rated on August 25 (Table 7). There were no significant differences between seed treatments for root rot on August 24. There were significant differences between seed treatments and cultivar for harvest moisture and test weight. There was no significant difference between seed treatments and cultivar for soybean yield.

TABLE 7. *Effect of Seed Treatment on Sudden Death Syndrome (SDS), Root Rot, and Yield of Soybean*

TREATMENT AND VARIETY[z]	SDS DI[y]	SDS DS[y]	SDS INDEX[y]	ROOT ROT %[x]	HARVEST MOISTURE %	TEST WEIGHT LB/BU	YIELD[w] BU/ACRE
Nontreated control, P25A27X	0.1	0.1	0.01	4.6	11.4 c	55.2 ab	52.5
ILeVO, P25A27X	0.0	0.0	0.00	3.9	11.5 bc	55.5 a	54.0
Saltro, P25A27X	0.0	0.0	0.00	4.0	11.6 b	55.2 ab	56.3
Nontreated control, P24T76E	0.0	0.0	0.00	5.3	11.9 a	55.2 ab	55.2
ILeVO, P24T76E	0.0	0.0	0.00	5.0	11.9 a	55.0 bc	53.9
Saltro, P24T76E	0.3	0.5	0.06	5.0	11.9 a	54.8 c	58.4
P-value[v]	*0.4890*	*0.4890*	*0.4890*	*0.3592*	*<.0001*	*0.0396*	*0.8113*

[z] Seed treatments were preapplied to the seed of varieties P25A27X (resistant) and P24T76E (susceptible).

[y] Sudden death syndrome (SDS) in each plot was rated for disease incidence (DI) and disease severity (DS) on August 25. DS refers to the percentage of plants with disease symptoms, and DS was rated using a 1–9 scale where 1 refers to low disease pressure and 9 refers to premature death of the plant. The SDS index was then calculated using the equation DX = (DI x DS)/9.

[x] Ten roots per plot were sampled from border rows at full pod (R4) growth stage, gently washed, and rated for root rot severity on a scale of 0–100% on August 25.

[w] Yields were adjusted to 13% moisture at harvest on October 14.

[v] Means followed by the same letter are not significantly different based on Fisher's Least Significant Difference test (LSD; α=0.05).

EVALUATION OF FUNGICIDES FOR FOLIAR DISEASES ON SOYBEAN IN CENTRAL INDIANA, 2020 (SOY20-21.ACRE)

D. E. P. Telenko, J. D. Ravellette, and S. Shim, Department of Botany and Plant Pathology, Purdue University West Lafayette, Indiana 47907-2054

SOYBEAN (*GLYCINE MAX* P35T75X)

Frogeye leaf spot, *Cercospora sojina*
Septoria brown spot, *Septoria glycines*
Cercospora leaf blight, *Cercospora kikuchii/ C. flagellaris*

A trial was established at the Purdue Agronomy Center for Research and Education (ACRE) in Tippecanoe County. The experiment was a randomized complete block design with four replications. Plots were 10 feet wide and 30 feet long and consisted of four rows, and the two center rows were used for evaluation. The previous crop was corn. Standard practices for soybean production in Indiana were followed. Soybean cultivar P35T75X was planted in 30-inch row spacing at a rate of 140,000 seeds/acre on June 2. All fungicide applications were applied at 15 gal/acre at 40 psi using a Lee self-propelled sprayer equipped with a 10-foot boom, fitted with six TJ-VS 8002 nozzles spaced 20 inches apart, at 3.6 mph. Fungicides were applied on July 29 at the beginning pod (R3) growth stage. Disease ratings were assessed on September 9 at the full seed (R6) growth stage. Frogeye leaf spot (FLS), Cercospora leaf blight (CLB), and Septoria brown spot (SBS) were rated for disease severity by visually assessing the percentage of symptomatic leaf area in the upper and lower canopies, respectively. The two center rows were harvested on October 16, and yields were adjusted to 13% moisture. Data were subjected to mixed model analysis of variance (SAS 9.4, 2019), and means were separated using Fisher's Least Significant Difference test (LSD; $\alpha=0.05$).

In 2020, weather conditions were not favorable for soybean disease. Frogeye leaf spot (FLS), Septoria brown spot (SBS), and Cercospora leaf blight (CLB) were the most prominent diseases in the trial and reached low severity. There were no differences between the nontreated control and all fungicide treatments for FLS on September 9 (Table 8). All fungicides reduced SBS on September 20 over the nontreated control. CLB was reduced by all fungicide treatments on September 20 over the nontreated control except Miravis Neo and Trivapro. No significant treatment effects were detected for soybean harvest moisture, test weight, and yield.

TABLE 8. *Effect of Fungicide on Foliar Disease Severity*

TREATMENT, RATE/ACRE[z]	FLS[y] %	SBS[y] %	CLB[y] %	HARVEST MOISTURE %	TEST WEIGHT LB/BU	YIELD[x] BU/ACRE
Nontreated check	0.05	7.5 a	0.5 ab	10.2	56.7	59.9
Miravis Neo 2.5 SE 13.7 fl oz	0.00	0.3 b	0.2 bc	10.0	56.2	63.5
Miravis Top 1.67 SC 13.7 fl oz	0.00	0.4 b	0.0 d	10.0	56.2	61.8
Miravis Neo 13.7 fl oz + Endigo ZCX 4.0 fl oz	0.00	0.5 b	0.4 a-c	10.1	10.1	56.1
Miravis Top 13.7 fl oz + Endigo ZCX 4.0 fl oz	0.00	0.4 b	0.1 cd	10.1	10.1	56.3
Delaro Complete 458 SC 8.0 fl oz	0.00	0.5 b	0.0 d	10.3	56.2	61.5
Lucento 4.17 SC 5.0 fl oz	0.00	0.9 b	0.0 d	10.3	56.4	60.3
Priaxor 4.17 SC 4.0 fl oz	0.00	0.8 b	0.1 cd	10.6	56.8	61.8
Revytek 3.33 LC 8.0 fl oz	0.03	0.9 b	0.0 d	10.3	56.6	57.2
Trivapro 2.21 SE 13.7 fl oz	0.00	0.9 b	0.6 a	10.1	56.2	65.4
P-value[w]	*0.0678*	*<.0001*	*0.0117*	*0.2859*	*0.5116*	*0.1542*

[z] Fungicide treatments were applied on July 29 at the beginning pod (R3) growth stage, and all treatments contained a nonionic surfactant (Preference) at a rate of 0.25% v/v except Delaro Complete 458 SC, which contained induce 0.12 % v/v.

[y] Foliar disease incidence was rated on a scale of 0–100% of plants within a plot with disease symptoms on September 9. FLS = frogeye leaf spot, SBS = Septoria brown spot, CLB = Cercospora leaf blight.

[x] Yields were adjusted to 13% moisture at harvest on October 16.

[w] Means followed by the same letter are not significantly different based on Fisher's Least Significant Difference (LSD; α=0.05).

EVALUATION OF FUNGICIDES FOR WHITE MOLD IN SOYBEAN IN CENTRAL INDIANA, 2020 (SOY20-31.ACRE)

S. Shim, J. D. Ravellette, and D. E. P. Telenko, Department of Botany and Plant Pathology, Purdue University West Lafayette, Indiana 47907-2054

SOYBEAN (*GLYCINE MAX* P35T75X)

White mold, *Sclerotinia sclerotiorum*

A trial was established at the Purdue Agronomy Center for Research and Education (ACRE) in Tippecanoe County, Indiana. The experiment was a randomized block design with four replications. Plots were 10 feet wide and 30 feet long and consisted of four rows, and the two center rows were used for evaluation. The previous crop was corn. Standard practices for soybean production in Indiana were used. Soybean cultivar P34A79X was planted in 30-inch row spacing at a rate of 10 seeds/foot on May 24. All fungicide applications were applied at 15 gal/acre at 40 psi using a Lee self-propelled sprayer equipped with a 10-foot boom, fitted with six TJ-VS 8002 nozzles spaced 20 inches apart, at 3.6 mph. Fungicides were applied on May 25 at 1 day after planting (DAP), June 2 at emergence/cotyledon (VE/VC), June 24 at third node/fourth node (V3/V4), July 1 at fifth node (V5), July 8 at beginning bloom (R1), July 15 full bloom (R2), and July 29 at beginning seed/full pod (R3/R4) growth stages. Injury was assessed as the percent of cupped and puckered plants per plot on July 17. Disease severity was assessed on September 15 at the R6 (full pod) growth stage. White mold was rated by visually scoring 30 random plants in each plot as 0 (no symptoms), 1 (only lateral branches with lesions), 2 (lesions on main stem infection but little to no effect on pod fill), or 3 (lesions on main stem resulting in poor pod fill or plant death), and then the white mold index (DSI) was calculated as DSI = 100 all plants 3, DSI = 0 all plants 0. The two center rows were harvested on October 14, and yields were adjusted to 13% moisture. Data were subjected to mixed model analysis of variance (SAS 9.4, 2020), and means were compared using Fisher's Least Significant Difference test (LSD; $\alpha=0.05$).

In 2020, weather conditions were unfavorable for soybean disease. White mold was present in the trial but only reached low levels. Injury severity was significantly increased for Cobra at all timing on July 17 (Table 9). There was no significant difference between fungicide treatments and nontreated control for white mold on September 15. Cobra plus Oxidate applied at V4/V5 and at R3 led to increased canopy yellowing on September 15 over the nontreated controls, although this treatment was not significantly different from the Cobra plus Endura program applied at V4/V5 and at R3 and the Oxidate program applied at R1, R2 and R3. There was no significant effect of treatments on soybean yield.

TABLE 9. *Effect of Fungicide on Foliar Disease Severity and Soybean Yield*

TREATMENT, RATE/ACRE, AND TIMING[z]	INJURY %[y]	WHITE MOLD INDEX[x]	YELLOW %[w]	HARVEST MOISTURE %	TEST WEIGHT LB/BU	YIELD[v] BU/A
Nontreated control	0.0 c	2.5	11.3 bcd	10.6	55.9	68.2
Endura 70 WDG 8.0 oz at R1 fb Endura 70 WDG 8.0 oz at R3	0.0 c	1.7	6.8 d-g	10.6	56.0	72.4
Lektivar 16.0 fl oz at R1 fb Lektivar 16.0 fl oz at R3	0.0 c	1.1	7.5 d-g	10.6	55.6	75.2
Cobra 6 fl oz + OxiDate 5.0, 1% v/v at V4-V5 fb Cobra 6.0 fl oz + OxiDate 5.0, 1% v/v at R3	3.3 b	3.3	16.3 a	10.6	55.9	69.7
Endura 70 WDG 6.0 oz + Priaxor Xemium 4.0 fl oz at R1 fb Endura 70 WDG 6.0 oz + Priaxor Xemium 4.0 fl oz at R3	0.0 c	1.4	3.0 g	10.6	55.7	70.6
Cobra 6.0 fl oz at R1	23.8 a	1.4	5.5 efg	10.6	55.9	69.9
Cobra 6.0 fl oz at V4/V5	4.0 b	1.1	7.5 d-g	10.6	55.9	72.0
Cobra 6.0 fl oz + Endura 70 WDG 8.0 oz at V4/V5 fb Cobra 6.0 fl oz + Endura 70 WDG 8.0 oz at R3	3.3 b	2.5	12.5 abc	10.6	55.4	70.3
Endura 70 WDG 8.0 oz at R3	0.0 c	0.8	6.3 efg	10.6	55.8	69.9
Xyway LFR 15.2 fl oz In-furrow	0.0 c	1.7	4.8 fg	10.5	56.1	71.2
Xyway LFR 15.2 fl oz Banded over row after planting	0.5 c	2.2	7.5 d-g	10.6	55.8	70.4
Lucento 4.17 SC 5.5 fl oz at V4/V5	0.0 c	1.4	10.0 b-e	10.6	55.8	72.0
Endura 70 WDG 8.0 oz broadcast after planting	0.0 c	1.9	8.8 c-f	10.6	55.8	75.0
Endura 70 WDG 8.0 oz at V2/V3	0.0 c	2.2	8.8 c-f	10.6	55.9	70.2
Endura 70 WDG 8.0 oz at R3 (drop nozzle)	0.0 c	2.8	6.3 efg	10.6	55.9	72.5
Lektivar 16.0 fl oz at R3 (drop nozzle)	0.0 c	1.1	10.0 b-e	10.6	56.0	75.0
OxiDate 5.0 1% v/v at R1 fb OxiDate 5.0 1% v/v at R2 fb OxiDate 5.0 1% v/v at R3	0.0 c	1.7	13.8 ab	10.6	55.7	73.4
Nontreated control	0.0 c	2.0	8.8 c-f	10.6	56.1	74.9
P-value[u]	<.0001	0.8283	0.0001	0.7088	0.8366	0.3868

[z] Fungicide treatments were applied on May 25, June 2, June 24, July 1, July 8, July 15, and July 29 at the 1 DAP, emergence/cotyledon (VE/VC), third/forth node (V3/V4), fifth node (V5), beginning bloom (R1), full bloom (R2), and beginning seed/full pod (R3/R4) growth stages. All plots were inoculated with *S. sclerotiorum*. All treatments contained a nonionic surfactant (Preference) at a rate of 0.25% v/v. fb = followed by.

[y] Injury was assessed as a percent of cupped and puckered plants per plot on July 17.

[x] White mold disease incidence (total DSI) was rated on September 15 by visually assessing the number of 30 random plants selected and scored in each plot. White mold index (DSI) was then calculated using the equation scores of 30 plants totaled for each plot/0.9, DSI = 100 all plants 30, DSI=0 all plants 0.

[w] Yellow was visually assessed as a percentage (0–100%) of crop canopy on September 15.

[v] Yields were adjusted to 13% moisture at harvest on October 14.

[u] Means followed by the same letter are not significantly different based on Fisher's Least Significant Difference test (LSD; α=0.05).

EVALUATION OF FUNGICIDES FOR WHITE MOLD IN SOYBEAN IN CENTRAL INDIANA, 2020 (SOY20-32.ACRE)

E. A. Duncan, S. Shim, J. D. Ravellette, and D. E. P. Telenko, Department of Botany and Plant Pathology, Purdue University West Lafayette, Indiana 47907-2054

SOYBEAN (*GLYCINE MAX* P35T75X)

White mold, *Sclerotinia sclerotiorum*

A trial was established at the Purdue Agronomy Center for Research and Education (ACRE) in Tippecanoe County, Indiana. The experiment was a randomized block design with four replications. Plots were 10 feet wide and 30 feet long and consisted of four rows, and the two center rows were used for evaluation. Standard practices for soybean production in Indiana were used. The previous crop was corn. The soybean cultivar P34A79X was planted in 30-inch row spacing at a rate of 100,000 seeds/acre and 160,000 seeds/acre on June 1. All plots were inoculated with *Sclerotinia sclerotiorum* at 1.25 g/foot within the seedbed at planting. Standard practices for nonirrigated soybean production in Indiana were followed. All fungicide applications were applied at 15 gal/acre at 40 psi using a Lee self-propelled sprayer equipped with a 10-foot boom, fitted with six TJ-VS 8002 nozzles spaced 20 inches apart, at 3.6 mph. Fungicides were applied on July 1, July 29, and August 13 at fourth trifoliolate/fifth trifoliolate (V4/V5), beginning pod (R3), and beginning seed (R5) growth stages, respectively. Disease severity was assessed on September 15 at full pod (R6) growth stage. White mold was rated by visually scoring 30 random plants in each plot as 0 (no symptoms), 1 (only lateral branches with lesions), 2 (lesions on main stem infection but little to no effect on pod-fill), or 3 (lesions on main stem resulting in poor pod fill or plant death), and then the white mold index (DSI) was calculated as DSI = 100 all plants 3, DSI = 0 all plants 0. The two center rows were harvested on October 16, and yields were adjusted to 13% moisture. Data were subjected to mixed model analysis of variance (SAS 9.4, 2020), and means were compared using Fisher's Least Significant Difference test (LSD; α=0.05).

In 2020, weather conditions were not favorable for soybean disease. White mold was present in the trial but only reached low levels. There was no significant difference between fungicide treatments and the nontreated control for white mold symptoms on September 15 (Table 10). There was no significant effect of treatment on moisture, test weight, or soybean yield.

TABLE 10. *Effect of Fungicide on Disease Severity and Soybean Yield*

SEEDING RATE, TREATMENT, RATE/ACRE, AND TIMING[z]	WHITE MOLD DSI INDEX[y]	HARVEST MOISTURE %	TEST WEIGHT LB/BU	YIELD[x] BU/ACRE
100,000 seeds/acre, Nontreated control	0.0	10.6	56.2	71.9
100,000 seeds/acre, Endura 70 WDG 8.0 oz at R3	0.0	10.6	56.6	74.7
100,000 seeds/acre, Endura 70 WDG 8.0 oz at R5	0.8	10.7	56.3	74.3
100,000 seeds/acre, Cobra 6.0 fl oz at V4/V5	0.0	10.8	56.4	73.2
160,000 seeds/acre, Nontreated control	0.0	10.6	56.2	75.6
160,000 seeds/acre, Endura 70 WDG 8.0 oz at R3	0.0	10.8	56.3	79.2
160,000 seeds/acre, Endura 70 WDG 8.0 oz at R5	0.3	10.6	56.4	72.0
160,000 seeds/acre, Cobra 6.0 fl oz at V4/V5	0.3	10.9	56.4	77.4
100,000 seeds/acre, Fertilizer, Nontreated control	0.3	10.7	56.5	72.9
100,000 seeds/acre, Fertilizer, Endura 70 WDG 8.0 oz at R3	0.3	10.5	56.4	68.4
100,000 seeds/acre, Fertilizer, Endura 70 WDG 8.0 oz at R5	0.0	10.6	56.3	71.8
100,000 seeds/acre, Fertilizer, Cobra 6.0 fl oz at V4/V5	0.8	10.6	56.4	69.5
160,000 seeds/acre, Fertilizer, Nontreated control	0.6	10.7	56.3	76.5
160,000 seeds/acre, Fertilizer, Endura 70 WDG 8.0 oz at R3	0.0	10.5	56.5	71.7
160,000 seeds/acre, Fertilizer, Endura 70 WDG 8.0 oz at R5	0.0	10.6	56.5	71.8
160,000 seeds/acre, Fertilizer, Cobra 6.0 fl oz at V4/V5	0.0	10.6	56.1	74.4
P-value[w]	0.7767	0.8015	0.8183	0.3786

[z] Fungicide treatments were applied on July 1 at fourth trifoliolate/fifth trifoliolate (V4/V5), July 29 at beginning pod (R3), and August 13 at beginning seed (R5) growth stages. All plots were inoculated with *S. sclerotiorum*. All treatments contained a nonionic surfactant (Preference) at a rate of 0.25% v/v.

[y] Disease severity was assessed on September 15 at the full seed (R6) growth stage. White mold was rated by visually scoring 30 random plants in each plot as 0 (no symptoms), 1 (only lateral branches with lesions), 2 (lesions on main stem infection but little to no effect on pod-fill), or 3 (lesions on main stem resulting in poor pod fill or plant death), and then the white mold index (DSI) was calculated as DSI = 100 all plants 3, DSI = 0 all plants 0.

[x] Yields were adjusted to 13% moisture at harvest on October 16.

[w] Means followed by the same letter are not significantly different based on Fisher's Least Significant Difference test (LSD; α=0.05).

FUSARIUM HEAD BLIGHT (FHB) UNIFORM FUNGICIDE TRIAL IN CENTRAL INDIANA, 2020 (WHT20-01.ACRE)

D. E. P. Telenko, J. D. Ravellette, and S. Shim, Department of Botany and Plant Pathology, Purdue University West Lafayette, Indiana 47907-2054

WHEAT (*TRITICUM AESTIVUM* P25R40)

Fusarium head blight, *Fusarium graminearum*
Stagnospora leaf and glume blotch, *Phaeosphaeria nodorum*

A trial was established at the Purdue Agronomy Center for Research and Education (ACRE) in Tippecanoe County, Indiana. The experiment was a randomized complete block design with four replications. Plots were 7.5 feet wide and 20 feet long and consisted of 12 rows spaced 7.5 inches apart, and the center of each plot was used for evaluation. The previous crop was corn. Prior to planting, the field was disked and chisel-plowed on October 9, 2019, and cultivated on October 10, 2019. Nitrogen MAP (11-52-0) at 300 lb/acre was applied on October 9, 2019, and nitrogen (28%) at 30 gal/acre was applied on March 7, 2020. On October 15, 2019, wheat cultivar P25R40 was drilled at 7.5-inch spacing. Harmony Extra at 0.8 oz/acre plus AMS at 2 lb/acre plus NIS at 0.25% v/v was applied on April 28, 2020, for weed management. All fungicide applications were applied at 15 gal/acre at 40 psi using a Lee self-propelled sprayer equipped with a 10-foot boom, fitted with six TJ-VS 8002 nozzles spaced 20 inches apart and directed forward and backward at a 45-degree angle at 3.6 mph. Fungicides were applied on May 21, at Feekes growth stage 10.3, May 29 at the Feekes growth stage 10.5.1, and June 3 at Feekes growth stage 10.5.3. All plots were inoculated with a mixture of isolates of *Fusarium graminearum* endemic to Indiana on May 29. The spore suspension (50,000 spores/ml) was applied at 300 ml/plot. Disease ratings were assessed on June 17. Fusarium head blight (FHB) disease incidence (DI) was measured as the number of infected heads out of 60 plants in each plot and calculated as a percentage. FHB disease severity (DS) was rated by visually assessing the percentage of the infected head. FHB index was calculated as (FHB % DI multiplied by average FHB % DS)/100 per plot. Disease severity of Stagnospora leaf and glume blotch was rated by visually assessing the percentage of symptomatic leaf tissue on five flag leaves per plot for leaf blotch and five heads per plot for glume blotch. Values for each plot were averaged before analysis. The eight center rows of each plot were harvested with a Kincaid plot combine on July 7, and yields were adjusted to 13.5% moisture. Data were subjected to mixed model analysis of variance (SAS 9.4, 2019), and means were compared using Fisher's Least Significant Difference test (LSD; α=0.05).

In 2020, weather conditions were moderately favorable for Fusarium head blight (FHB), leaf blotch, and glume blotch diseases. FHB was the most prominent disease. FHB % DI, FHB % DS, and FHB index were reduced by all fungicide treatments over the nontreated control on June 17 (Table 11). Miravis Ace applied at Feekes 10.3 resulted in the lowest FHB % DI and FHB index but was not significantly different from all other fungicide treatments or timings. All fungicides and timings reduced percent incidence of leaf blotch over the nontreated control. There were no differences in treatments from nontreated control for percent glume blotch and percent Fusarium damaged kernels (FDK) (Tables 11 and 12). The concentration of deoxynivalenol (DON) was significantly reduced over the nontreated control by all treatments, while Caramba applied at 10.5.1 had higher levels of DON than fungicide programs of Miravis Ace applied at 10.5.1, 10.5.3, and Miravis Ace at 10.5.1 followed by Prosaro or Caramba at 10.5.3 (Table 12). There was no difference in wheat moisture, test weight, or yield.

TABLE 11. *Effect of Fungicide on Fusarium Head Blight and Foliar Diseases in Wheat*

TREATMENT, RATE/ACRE, TIMING[z]	FHB % DI[y]	FHB % DS[y]	FHB INDEX[x]	LEAF BLOTCH[w] %	GLUME BLOTCH[w] %
Nontreated control	65.0 a	18.3 a	11.6 a	8.7 a	0.4
Prosaro 421 SC 6.5 fl oz at 10.5.1	29.5 bc	11.1 b	3.3 bc	4.9 b	0.1
Caramba 90 EC 13.5 fl oz at 10.5.1	37.1 bc	11.1 b	4.5 bc	2.8 b	0.4
Sphaerex 7.3 fl oz at 10.5.1	32.1 bc	10.6 b	3.6 bc	3.6 b	0.4
Prosaro Pro 10.3 fl oz at 10.5.1	32.5 bc	11.1 b	3.9 bc	2.4 b	0.4
Miravis Ace 5.2 SC 13.7 fl oz at 10.3	23.8 c	7.5 b	1.8 c	2.9 b	0.0
Miravis Ace 5.2 SC 13.7 fl oz at 10.5.1	25.0 bc	10.9 b	2.7 bc	3.7 b	0.4
Miravis Ace 5.2 SC 13.7 fl oz at 10.5.3	34.6 bc	9.2 b	3.4 bc	3.3 b	0.0
Miravis Ace 5.2 SC 13.7 fl oz at 10.5.1 Prosaro 421 SC 6.5 fl oz at 10.5.3	30.0 bc	10.9 b	3.3 bc	3.3 b	0.0
Miravis Ace 5.2 13.7 fl oz at 10.5.1 fb Caramba 90 EC 13.5 fl oz at 10.5.3	26.7 bc	8.8 b	2.4 bc	2.9 b	0.4
Miravis Ace 5.2 13.7 fl oz at 10.5.1 fb Folicur 4.0 fl oz @ 10.5.3	25.4 bc	7.9 b	2.0 c	3.1 b	0.8
P-value[v]	<.0001	0.0006	<.0001	0.0070	0.7876

[z] Fungicide treatments were applied at Feekes 10.3, 10.5.1, and 10.5.3. All treatments contained a nonionic surfactant (Preference) at a rate of 0.125% v/v. All plots were inoculated with *Fusarium graminearum* spore suspension (40,000–100,000 spores/ml) after the treatment at Feekes 10.5.1. Spore suspension was applied at 300 ml/plot on May 29. fb = followed by.

[y] FHB % DI = Fusarium head blight (FHB) percent disease incidence was measured as the number of infected heads out of 60 plants in each plot and calculated as a percentage. FHB % DS = FHB percent disease severity was rated by visually assessing the severity of the infected heads.

[x] FHB index was calculated as (FHB % DI/average FHB % DS)/100 per plot.

[w] Disease severity of Stagnospora leaf and glume blotch was rated by visually assessing the percentage of symptomatic leaf tissue on five flag leaves per plot for leaf blotch and five heads per plot for glume blotch.

[v] Means followed by the same letter are not significantly different based on Fisher's Least Significant Difference test (LSD; α=0.05).

TABLE 12. *Effect of Fungicide on Deoxynivalenol (DON), Fusarium Damaged Kernels (FDK), and Yield of Wheat*

TREATMENT, RATE/ACRE, TIMING[z]	DON[y] PPM	FDK[x] %	HARVEST MOISTURE %	TEST WEIGHT LB/BU	YIELD[w] BU/A
Nontreated control	0.548 a	5.1	13.0	58.8	96.0
Prosaro 421 SC 6.5 fl oz at 10.5.1	0.209 bc	4.9	13.6	58.4	92.1
Caramba 90 EC 13.5 fl oz at 10.5.1	0.363 b	4.8	13.7	58.3	91.0
Sphaerex 7.3 fl oz at 10.5.1	0.205 bc	5.0	13.3	58.3	99.5
Prosaro PRO 400 SC 10.3 fl oz at 10.5.1	0.205 bc	5.0	12.9	59.1	98.5
Miravis Ace 5.2 SC 13.7 fl oz at 10.3	0.278 bc	5.5	13.7	58.6	95.4
Miravis Ace 5.2 SC 13.7 fl oz at 10.5.1	0.188 c	4.3	13.4	59.2	99.0
Miravis Ace 5.2 SC 13.7 fl oz at 10.5.3	0.150 c	5.3	14.0	58.7	94.4
Miravis Ace 5.2 SC 13.7 fl oz at 10.5.1 fb Prosaro 421 SC 6.5 fl oz at 10.5.3	0.158 c	4.9	13.7	58.9	98.3
Miravis Ace 5.2 13.7 fl oz at 10.5.1 fb Caramba 90 EC 13.5 fl oz at 10.5.3	0.148 c	5.5	13.7	58.5	96.1
Miravis Ace 5.2 13.7 fl oz at 10.5.1 fb Folicur 4.0 fl oz at 10.5.3	0.210 bc	4.9	13.6	58.6	98.0
P-value[v]	*0.0013*	*0.1775*	*0.0531*	*0.1928*	*0.4282*

[z] Fungicide treatments were applied at Feekes 10.3, 10.5.1, and 10.5.3. All treatments contained a nonionic surfactant (Preference) at a rate of 0.125% v/v. All plots inoculated with *Fusarium graminearum* spore suspension (40,000–100,000 spores/ml) after the treatment at Feekes 10.5.1. Spore suspension was applied at 300 ml/plot on May 29. fb = followed by.

[y] Analysis of the mycotoxin deoxynivalenol (DON) was completed by the University of Minnesota DON Testing Lab.

[x] FDK = percentage of Fusarium damaged kernels.

[w] Yields were adjusted to 13.5% moisture at harvest on July 7.

[v] Means followed by the same letter are not significantly different based on Fisher's Least Significant Difference test (LSD; α=0.05).

FUSARIUM HEAD BLIGHT (FHB) INTEGRATED MANAGEMENT TRIAL IN CENTRAL INDIANA, 2020 (WHT20-02.ACRE)

D. E. P. Telenko, J. D. Ravellette, and S. Shim, Department of Botany and Plant Pathology, Purdue University West Lafayette, Indiana 47907-2054

WHEAT (*TRITICUM AESTIVUM* P25R40 AND P25R61)

Fusarium head blight, *Fusarium graminearum*
Stagnospora leaf and glume blotch, *Phaeosphaeria nodorum*

A research trial was established at the Purdue Agronomy Center for Research and Education (ACRE) in Tippecanoe County, Indiana. The experiment was a randomized complete block design with four replications. Plots were 7.5 feet wide and 20 feet long and consisted of 12 rows spaced 7.5 inches apart, and the center of each plot was used for evaluation. The previous crop was corn. Prior to planting, the field was disked and chisel-plowed on October 9, 2019, and cultivated on October 10, 2019. Nitrogen MAP (11-52-0) at 300 lb/acre was applied on October 9, 2019, and nitrogen (28%) at 30 gal/acre was applied on March 7, 2020. On October 15, 2019, wheat cultivars P25R40 (scab susceptible) and P25R61 (scab moderately resistant) were drilled at 7.5-inch spacing. Harmony Extra at 0.8 oz/acre plus AMS at 2 lb/acre plus NIS at 0.25% v/v was applied on April 28, 2020, for weed management. All fungicide applications were applied at 15 gal/acre and 40 psi using a Lee self-propelled sprayer equipped with a 10-foot boom, fitted with six TJ-VS 8002 nozzles spaced 20 inches apart and directed forward and backward at a 45-degree angle at 3.6 mph. Fungicides were applied on May 21 at Feekes growth stage 10.3, May 29 at Feekes growth stage 10.5.1, and June 3 at Feekes growth stage 10.5.3. All plots were inoculated with a mixture of isolates of *Fusarium graminearum* endemic to Indiana on May 29 except for the nontreated, noninoculated control. The spore suspension (50,000 spores/ml) was applied at 300 ml/plot. Disease ratings were assessed on June 17. Fusarium head blight (FHB) disease incidence (DI) was measured as the number of infected heads out of 60 plants in each plot and calculated as a percentage. FHB disease severity (DS) was rated by visually assessing the percentage of the infected head. FHB index was calculated as (FHB % DI multiplied by average FHB % DS)/100 per plot. Disease severity of Stagnospora leaf blotch was rated by visually assessing the percentage of symptomatic leaf tissue on five flag leaves per plot. Values for each plot were averaged before analysis. The eight center rows of each plot were harvested with a Kincaid plot combine on July 7, and yields were adjusted to 13.5% moisture. Data were subjected to mixed model analysis of variance (SAS 9.4, 2019), and means were compared using Tukey-HSD (α=0.05).

In 2020, weather conditions were moderately favorable for Fusarium head blight (FHB) and leaf blotch diseases. Fusarium head blight was the most prominent disease. Main effects of cultivar and fungicide treatment are presented. FHB % DI, FHB %DS, FHB index, and leaf blotch were lowest in the moderately resistant cultivar P25R61 (Table 13). FHB % DI and FHB index were reduced by all fungicide treatments over the nontreated inoculated control on June 17. Only treatments of Prosaro and Miravis Ace applied at Feekes 10.3 resulted in the lowest FHB % DS, and leaf blotch was reduced by the Miravis Ace applied at Feekes 10.3. The concentration of deoxynivalenol (DON) was significantly reduced over the nontreated, inoculated control by Miravis Ace applied at Feekes 10.5.1, but this was not different from the nontreated, noninoculated control

(Table 14). There were no significant differences in percent of Fusarium damaged kernels (FDK). Wheat test weight and yield were highest in the cultivar P25R40. Miravis Ace applied at Feekes 10.3 had significantly higher test weight than both nontreated controls, while Prosaro applied at Feekes 10.5.1 increased yield over the nontreated, non inoculated control but was not different from the other fungicide programs or nontreated, inoculated control.

TABLE 13. *Effect of Cultivar and Fungicide on Fusarium Head Blight (FHB) and Foliar Disease in Wheat*

	FHB % DI [Y]	FHB % DS[Y]	FHB INDEX[X]	LEAF BLOTCH[W] %
Cultivar				
P25R40	41.0 a[V]	14.8 a	6.5 a	5.7 a
P25R61	14.2 b	8.4 b	1.2 b	3.7 b
Fungicide program				
Nontreated control, inoculated control	41.9 a	15.8 a	8.1 a	7.7 a
Prosaro 421 SC 6.5 fl oz at 10.5.1	29.6 b	9.3 b	3.2 b	2.6 b
Miravis Ace 5.2 SC 13.7 fl oz at 10.5.1	20.4 b	9.0 b	2.1 b	4.9 ab
Miravis Ace 5.2 SC 13.7 fl oz at 10.3	21.5 b	11.0 ab	2.5 b	4.2 ab
Miravis Ace 5.2 SC 13.7 fl oz at 10.5.1 fb Folicur 4.0 fl oz at 10.5.3	26.3 b	10.2 ab	2.9 b	4.4 ab
Nontreated, noninoculated control	26.3 b	14.4 ab	4.4 b	4.3 ab
Cultivar P-value	<.0001	<.0001	<.0001	0.0156
Treatment P-value	<.0001	0.0060	<.0001	0.0259
Cul*Trt P-value	0.0147	0.0280	<.0001	0.1179

[Z] Fungicide treatments were applied at Feekes 10.3, 10.5.1, and 10.5.3. All treatments contained a nonionic surfactant (Preference) at a rate of 0.125% v/v. All plots were inoculated with *Fusarium graminearum* spore suspension (50,000 spores/ml) after the treatment at Feekes 10.5.1 except for the nontreated, noninoculated control. Spore suspension was applied at 300 ml/plot on May 29. fb = followed by.

[Y] FHB % DI = Fusarium head blight (FHB) percent disease incidence was measured as the number of infected heads out of 60 plants in each plot and calculated as a percentage. FHB % DS= FHB percent disease severity was rated by visually assessing the severity of the infected heads.

[X] FHB index was calculated as (FHB % DI/average FHB % DS)/100 per plot.

[W] Disease severity of Stagnospora leaf and glume blotch was rated by visually assessing the percentage of symptomatic leaf tissue on five flag leaves per plot for leaf blotch and five heads per plot for glume blotch.

[V] Means followed by the same letter are not significantly different based on Tukey-HSD (α=0.05).

TABLE 14. *Effect of Cultivar and Fungicide on Deoxynivalenol (DON), Fusarium Damaged Kernels (FDK), and Yield of Wheat*

	DON[y] PPM	FDK[x] %	HARVEST MOISTURE %	TEST WEIGHT LB/BU	YIELD[w] BU/A
Cultivar					
P25R40	0.212	6.0	13.3	59.0 a	98.3 a[v]
P25R61	0.248	6.2	13.2	57.5 b	93.2 b
Fungicide program					
Nontreated control, inoculated control	0.335 ab	6.6	12.9 b	58.0 bc	94.4
Prosaro 421 SC 6.5 fl oz at 10.5.1	0.263 ab	6.1	13.1 ab	58.6 ab	100.1
Miravis Ace 5.2 SC 13.7 fl oz at 10.5.1	0.089 b	5.8	13.5 a	58.4 abc	96.4
Miravis Ace 5.2 SC 13.7 fl oz at 10.3	0.373 a	5.6	13.2 a	58.7 a	97.8
Miravis Ace 5.2 SC 13.7 fl oz at 10.5.1 fb Folicur 4.0 fl oz at 10.5.3	0.240 ab	6.1	13.5 ab	58.3 abc	96.0
Nontreated, noninoculated control	0.083 b	6.4	13.2 a	57.8 c	89.8 b
Cultivar P-value	*0.5047*	*0.3700*	*0.0715*	*<.0001*	*0.0071*
Treatment P-value	*0.0147*	*0.1657*	*0.0138*	*0.0032*	*0.0508*
*Cul*Trt P-value*	*0.8107*	*0.8792*	*0.5677*	*0.6101*	*0.8113*

[z] Fungicide treatments were applied at Feekes 10.3, 10.5.1, and 10.5.3. All treatments contained a nonionic surfactant (Preference) at a rate of 0.125% v/v. All plots were inoculated with *Fusarium graminearum* spore suspension (50,000 spores/ml) after the treatment at Feekes 10.5.1 except for the nontreated, noninoculated control. Spore suspension was applied at 300 ml/plot on May 29. fb = followed by.

[y] Analysis of the mycotoxin deoxynivalenol (DON) was completed by the University of Minnesota DON Testing Lab.

[x] FDK = percentage of Fusarium damaged kernels.

[w] Yields were adjusted to 13.5% moisture at harvest on July 7.

[v] Means followed by the same letter are not significantly different based on Tukey-HSD (α=0.05).

EVALUATION OF FOLIAR FUNGICIDES FOR WHEAT DISEASE MANAGEMENT IN CENTRAL INDIANA, 2020A (WHT20-05A.ACRE)

D. E. P. Telenko, J. D. Ravellette, and S. Shim, Department of Botany and Plant Pathology, Purdue University West Lafayette, Indiana 47907-2054

WHEAT (*TRITICUM AESTIVUM* P25R40)

Fusarium head blight, *Fusarium graminearum*
Stagnospora leaf and glume blotch, *Phaeosphaeria nodorum*

A field trial was established at the Purdue Agronomy Center for Research and Education (ACRE) in Tippecanoe County, Indiana. The experiment was a randomized complete block design with four replications. Plots were 7.5 feet wide and 20 feet long and consisted of 12 rows spaced 7.5 inches apart, and the center of each plot was used for evaluation. The previous crop was corn. Prior to planting, the field was disked and chisel-plowed on October 9, 2019, and cultivated on October 10, 2019. Nitrogen MAP (11-52-0) at 300 lb/acre was applied on October 9, 2019, and nitrogen (28%) at 30 gal/acre was applied on March 7, 2020. On October 15, 2019, wheat cultivar P25R40 was drilled at 7.5-inch spacing. Harmony Extra at 0.8 oz/acre plus AMS at 2 lb/acre plus NIS at 0.25% v/v was applied on April 28, 2020, for weed management. All fungicide applications were applied at 15 gal/acre and 40 psi using a Lee self-propelled sprayer equipped with a 10-foot boom, fitted with six TJ-VS 8002 nozzles spaced 20 inches apart and directed forward and backward at a 45-degree angle at 3.6 mph. Fungicides were applied on May 29, 2020, at the Feekes growth stage 10.5.1. All plots were inoculated with a mixture of isolates of *Fusarium graminearum* endemic to Indiana on May 29. The spore suspension (50,000 spores/ml) was applied at 300 ml/plot with the CO_2 handheld sprayer. Disease ratings were assessed on June 16. Fusarium head blight (FHB) disease incidence (DI) was measured as the number of infected heads out of 60 plants in each plot and calculated as a percentage. FHB disease severity (DS) was rated by visually assessing the percentage of the infected head. FHB index was calculated as (FHB % DI multiplied by average FHB % DS)/100 per plot. Disease severity of Stagnospora leaf and glume blotch was rated by visually assessing the percentage of symptomatic leaf tissue on five flag leaves per plot for leaf blotch and five heads per plot for glume blotch. Values for each plot were averaged before analysis. The eight center rows of each plot were harvested with a Kincaid plot combine on July 7, and yields were adjusted to 13.5% moisture. Data were subjected to mixed model analysis of variance (SAS 9.4, 2019), and means were compared using Fisher's Least Significant Difference test (LSD; α=0.05).

In 2020, weather conditions were moderately favorable for Fusarium head blight (FHB), leaf blotch, and glume blotch diseases. FHB was the most prominent disease. FHB % DI was not reduced by any fungicide over the nontreated control on June 16 (Table 15). FHB % DS was significantly reduced by all fungicide treatments over the nontreated control except for Prosaro. There were no differences in FHB index, % leaf blotch, or % glume blotch. The concentration of deoxynivalenol (DON) was not significantly different from the nontreated control for all treatments (Table 16). There was no significant difference in wheat moisture, test weight, or yield.

TABLE 15. *Effect of Fungicide on Fusarium Head Blight and Foliar Disease in Wheat*

TREATMENT AND RATE/ACRE[z]	FHB % DI[y]	FHB % DS[y]	FHB INDEX[x]	LEAF BLOTCH[w] %	GLUME BLOTCH[w] %
Nontreated control	22.5	19.6 a	4.3	6.0	4.6
Prosaro 422 SC 8.2 fl oz	28.3	13.6 ab	3.9	3.4	2.5
Miravis Ace 275 SC 13.7 fl oz	21.3	11.2 b	2.3	4.7	1.7
Prosaro PRO 400 SC 400 SC 10.3 fl oz	28.8	8.2 b	2.3	4.8	2.5
Caramba 90 EC 13.5 fl oz	30.8	10.8 b	3.4	4.6	0.8
BAS 84999F 7.3 fl oz	27.1	9.7 b	2.8	3.5	2.5
P-value[v]	*0.1416*	*0.0146*	*0.2724*	*0.6647*	*0.3438*

[z] Fungicide treatments were applied at Feekes 10.3, 10.5.1, and 10.5.3. All treatments contained a nonionic surfactant (Preference) at a rate of 0.125% v/v. All plots were inoculated with *Fusarium graminearum* spore suspension (40,000–100,000 spores/ml) after the treatment at Feekes 10.5.1. Spore suspension applied at 300 ml/plot on May 29.

[y] FHB % DI = Fusarium head blight (FHB) percent disease incidence was measured as the number of infected heads out of 60 plants in each plot and calculated as a percentage. FHB % DS = FHB percent disease severity was rated by visually assessing the severity of the infected heads.

[x] FHB index was calculated as (FHB % DI/average FHB % DS)/100 per plot.

[w] Disease severity of Stagnospora leaf and glume blotch was rated by visually assessing the percentage of symptomatic leaf tissue on five flag leaves per plot for leaf blotch and five heads per plot for glume blotch.

[v] Means followed by the same letter are not significantly different based on Fisher's Least Significant Difference test (LSD; α=0.05).

TABLE 16. *Effect of Fungicide on Deoxynivalenol (DON), Fusarium Damaged Kernels (FDK), and Yield in Wheat*

TREATMENT AND RATE/ACRE	DON[y] PPM	FDK[x] %	HARVEST MOISTURE %	TEST WEIGHT LB/BU	YIELD[w] BU/A
Nontreated control	0.3	4.0	12.4	58.3	94.6
Prosaro 422 SC 8.2 fl oz	0.2	4.0	12.5	58.6	92.7
Miravis Ace 275 SC 13.7 fl oz	0.2	3.9	12.9	58.8	94.5
Prosaro PRO 400 SC 400 SC 10.3 fl oz	0.2	3.9	12.5	58.5	99.5
Caramba 90 EC 13.5 fl oz	0.3	4.3	12.9	58.1	91.4
BAS 84999F 7.3 fl oz	0.1	4.1	12.3	58.1	97.5
P-value[v]	*0.4000*	*0.9873*	*0.4014*	*0.5703*	*0.4595*

[z] Fungicide treatments were applied at Feekes 10.3, 10.5.1, and 10.5.3. All treatments contained a nonionic surfactant (Preference) at a rate of 0.125% v/v. All plots were inoculated with *Fusarium graminearum* spore suspension (50,000 spores/ml) after the treatment at Feekes 10.5.1 except for the nontreated, noninoculated control. Spore suspension was applied at 300 ml/plot on May 29.

[y] Analysis of the mycotoxin deoxynivalenol (DON) was completed by the University of Minnesota DON Testing Lab.

[x] FDK = percentage of Fusarium damaged kernels.

[w] Yields were adjusted to 13.5% moisture at harvest on July 7.

[v] Means followed by the same letter are not significantly different based on Fisher's Least Significant Difference test (LSD; α=0.05).

EVALUATION OF FOLIAR FUNGICIDES FOR WHEAT DISEASE MANAGEMENT IN CENTRAL INDIANA, 2020B (WHT20-05B.ACRE)

D. E. P. Telenko, J. D. Ravellette, and S. Shim, Department of Botany and Plant Pathology, Purdue University West Lafayette, Indiana 47907-2054

WHEAT (*TRITICUM AESTIVUM* P25R40)

Fusarium head blight, *Fusarium graminearum*
Stagnospora leaf and glume blotch, *Phaeosphaeria nodorum*

A trial was established at the Purdue Agronomy Center for Research and Education (ACRE) in Tippecanoe County, Indiana. The experiment was a randomized complete block design with four replications. Plots were 7.5 feet wide and 20 feet long and consisted of 12 rows spaced 7.5 inches apart, and the center of each plot was used for evaluation. The previous crop was corn. Prior to planting, the field was disked and chisel-plowed on October 9, 2019, and cultivated on October 10, 2019. Nitrogen MAP (11-52-0) at 300 lb/acre was applied on October 9, 2019, and nitrogen (28%) at 30 gal/acre was applied on March 7, 2020. On October 15, 2019, wheat cultivar P25R40 was drilled at 7.5-inch spacing. Harmony Extra at 0.8 oz/acre plus AMS at 2 lb/acre plus NIS at 0.25% v/v was applied on April 28, 2020, for weed management. All fungicide applications were applied at 15 gal/acre and 40 psi using a Lee self-propelled sprayer equipped with a 10-foot boom, fitted with six TJ-VS 8002 nozzles spaced 20 inches apart and directed forward and backward at a 45-degree angle at 3.6 mph. Fungicides were applied on May 29, 2020, at Feekes growth stage 10.5.1. Disease ratings were assessed on June 16. Fusarium head blight (FHB) disease incidence (DI) was measured as the number of infected heads out of 60 plants in each plot and calculated as a percentage. FHB disease severity (DS) was rated by visually assessing the percentage of the infected head. FHB index was calculated as (FHB % DI multiplied by average FHB % DS)/100 per plot. Disease severity of Stagnospora leaf and glume blotch was rated by visually assessing the percentage of symptomatic leaf tissue on five flag leaves per plot for leaf blotch and five heads per plot for glume blotch. Values for each plot were averaged before analysis. The eight center rows of each plot were harvested with a Kincaid plot combine on July 7, and yields were adjusted to 13.5% moisture. Data were subjected to mixed model analysis of variance (SAS 9.4, 2019), and means were compared using Fisher's Least Significant Difference test (LSD; α=0.05).

In 2020, weather conditions were moderately favorable for Fusarium head blight (FHB), leaf blotch, and glume blotch diseases. FHB was the most prominent disease. FHB % DI and FHB index were reduced by Badge SC at both 1.0 and 1.8 pt as compared to the nontreated control (Table 17). There were no differences in treatments for FHB % DS, % leaf blotch, and % glume blotch. The concentration of deoxynivalenol (DON) was not significantly reduced over the nontreated control for all treatments (Table 18). There was no difference in wheat moisture, test weight, or yield.

TABLE 17. *Effect of Fungicide on Fusarium Head Blight and Foliar Disease in Wheat*

TREATMENT AND RATE/ACRE[z]	FHB % DI[y]	FHB % DS[y]	FHB INDEX[x]	LEAF BLOTCH %[w]	GLUME BLOTCH %[w]
Nontreated control	49.6a	24.0	11.6a	9.6	7.5
Badge SC 1.0 pt	25.4b	19.6	4.9b	9.0	2.5
Badge SC 1.8 pt	26.3b	19.9	5.1b	4.8	6.3
P-value[u]	0.0042	0.6598	0.0021	0.2461	0.1282

[z] Fungicide treatments were applied at Feekes 10.5.1. All treatments contained a nonionic surfactant (Preference) at a rate of 0.125% v/v on May 29.

[y] FHB % DI = Fusarium head blight (FHB) percent disease incidence was measured as the number of infected heads out of 60 plants in each plot and calculated as a percentage. FHB % DS = FHB percent disease severity was rated by visually assessing the severity of the infected heads.

[x] FHB index was calculated as (FHB % DI/average FHB % DS)/100 per plot.

[w] Disease severity of Stagnospora leaf and glume blotch was rated by visually assessing the percentage of symptomatic leaf tissue on five flag leaves per plot for leaf blotch and five heads per plot for glume blotch on June 16.

[u] Means followed by the same letter are not significantly different based on Fisher's Least Significant Difference test (LSD; α=0.05).

TABLE 18. *Effect of Fungicide on Deoxynivalenol (DON), Fusarium Damaged Kernels (FDK), and Yield in Wheat*

TREATMENT AND RATE/ACRE[z]	DON[y] PPM	FDK[x] %	HARVEST MOISTURE %	TEST WEIGHT LB/BU	YIELD[w] BU/ACRE
Nontreated control	0.5	5.0	12.2	58.3	96.6
Badge SC 1.0 pt	0.7	4.4	12.5	59.1	92.2
Badge SC 1.8 pt	0.5	5.0	12.2	58.4	95.9
P-value[v]	0.4067	0.5997	0.4384	0.5730	0.3680

[z] Fungicide treatments were applied at Feekes 10.5.1. All treatments contained a nonionic surfactant (Preference) at a rate of 0.125% v/v.

[y] Analysis of the mycotoxin deoxynivalenol (DON) was completed by the University of Minnesota DON Testing Lab.

[x] FDK = percentage of Fusarium damaged kernels.

[w] Yields were adjusted to 13.5% moisture at harvest on July 7.

[v] Means followed by the same letter are not significantly different based on Fisher's Least Significant Difference test (LSD; α=0.05).

PINNEY PURDUE AGRICULTURAL CENTER (PPAC)

UNIFORM FUNGICIDE COMPARISON FOR TAR SPOT OF CORN IN NORTHWESTERN INDIANA, 2020 (COR20-03.PPAC)

T. J. Ross, J. D. Ravellette, S. Shim, and D. E. P. Telenko, Department of Botany and Plant Pathology, Purdue University West Lafayette, Indiana 47907-2054

CORN (*ZEA MAYS* W2585SSRIB)

Tar spot, *Phyllachora maydis*

A trial was established at the Pinney Purdue Agricultural Center (PPAC) in Porter County, Indiana. The experiment was a randomized complete block design with four replications. Plots were 10 feet wide and 30 feet long and consisted of four rows, and the two center rows were used for evaluation. The previous crop was corn. Standard practices for grain corn production in Indiana were followed. Corn hybrid W2585SSRIB was planted in 30-inch row spacing at a rate of 34,000 seeds/acre on June 8. All fungicide applications were applied at 15 gal/acre and 40 psi using a Lee self-propelled sprayer equipped with a 10-foot boom, fitted with six TJ-VS 8002 nozzles spaced 20 inches apart, at 3.6 mph. Fungicides were applied on August 7 at the tassel/silk (VT/R1) growth stage. Disease ratings were assessed on September 22 at the dent (R5) and on October 13 at the maturity (R6) growth stages. Disease severity was rated by visually assessing the percentage of symptomatic leaf area on five plants in each plot at the ear leaf (EL), ear leaf minus 2 leaves (EL–2), and ear leaf plus 2 leaves (EL+2). The values of the five leaves for each plot were averaged before analysis. The two center rows of each plot were harvested on November 6, and yields were adjusted to 15.5% moisture. Data were subjected to mixed model analysis of variance (SAS 9.4, 2019), and means were compared using Fisher's Least Significant Difference test (LSD; α=0.05).

In 2020, weather conditions were favorable for disease. Tar spot was the most prominent diseases in the trial and reached moderate to high severity. All fungicides significantly reduced the percentage of stromata and the percentage of chlorotic and necrotic symptoms of tar spot on the EL–2, EL, and EL+2 over the non-treated control on September 22 (Table 19). Aproach Prima and Miravis Neo had the lowest percentage of stromata on the EL–2 on September 22 but was not significantly different from Revytek, Veltyma, Headline

AMP, and Delaro. All fungicides significantly reduced the percentage of stromata and chlorotic and necrotic symptoms on the EL-2, EL, and EL+2 as compared to the nontreated control on October 13 (Table 20). All fungicide treatments significantly increased the percentage of green canopy over the nontreated control on October 13 (Table 21). Veltyma had the highest percentage of green canopy but was not significantly different from Revytek, Aproach Prima, Miravis, and Delaro. There were no significant differences between treatments for test weight and yield of corn.

TABLE 19. *Effect of Fungicide on Tar Spot on September 22*

TREATMENT AND RATE/ACRE[z]	TAR SPOT % STROMA[y] EL-2	TAR SPOT % STROMA[y] EL	TAR SPOT % STROMA[y] EL+2	TAR SPOT % CHLOR/NEC[x] EL-2	TAR SPOT % CHLOR/NEC[x] EL	TAR SPOT % CHLOR/NEC[x] EL+2
Nontreated control	21.0 a	15.0 a	10.8 a	5.7 a	1.9 a	0.9 a
Revytek 3.33 LC 8.0 fl oz	8.3 cde	5.1 cde	4.8 cd	0.2 b	0.1 c	0.0 b
Veltyma 3.34 S 7.0 fl oz	9.8 b-e	5.1 cde	3.5 d	1.0 b	0.1 c	0.1 b
Headline 2.08 SC 6.0 fl oz	11.0 bcd	6.3 b-e	5.1 cd	1.2 b	0.3 b	0.1 b
Headline AMP 1.68 SE 10.0 fl oz	9.1 b-e	7.3 bcd	5.0 cd	0.6 b	0.2 c	0.1 b
Aproach Prima 2.34 SC 6.8 fl oz	4.9 e	4.4 de	4.6 cd	0.2 b	0.0 c	0.1 b
Miravis Neo 2.5 SE 13.7 fl oz	4.6 e	4.3 de	6.1 c	0.0 b	0.0 c	0.0 b
Delaro 325 SC 8.0 fl oz	6.7 de	4.0 e	3.7 d	0.4 b	0.1 c	0.0 b
Lucento 4.17 SC 5.0 fl oz	14.0 b	9.3 b	6.9 bc	2.4 b	0.9 b	0.3 b
Tilt 3.6 EC 4.0 fl oz	12.5 bc	8.0 bc	9.1 ab	0.4 b	0.1 c	0.1 b
P-value[w]	<.0001	<.0001	<.0001	0.0096	0.0003	0.0002

[z] Fungicide treatments were applied on August 7 at the tassel/silk (VT/R1) growth stage, and all treatments contained a nonionic surfactant (Preference) at a rate of 0.25% v/v.

[y] Tar spot stromata was visually assessed as a percentage (0–100%) of leaf area on five plants in each plot at the ear leaf (EL), ear leaf minus two (EL-2), and ear leaf plus two (EL+2).

[x] Tar spot chlorosis and necrosis symptoms were visually assessed as a percentage (0–100%) of leaf area on five plants in each plot at the ear leaf (EL), ear leaf minus two (EL-2), and ear leaf plus two (EL+2).

[w] Means followed by the same letter are not significantly different based on Fisher's Least Significant Difference test (LSD; α=0.05).

TABLE 20. *Effect of Fungicide on Tar Spot on October 13*

TREATMENT AND RATE/ACRE[z]	TAR SPOT % STROMA[y] EL–2	TAR SPOT % STROMA[y] EL	TAR SPOT % STROMA[y] EL+2	TAR SPOT % CHLOR/ NEC[x] EL–2	TAR SPOT % CHLOR/ NEC[x] EL	TAR SPOT % CHLOR/ NEC[x] EL+2
Nontreated control	35.0 a	32.0 a	25.0 a	96.8 a	78.5 a	50.5 a
Revytek 3.33 LC 8.0 fl oz	5.7 d	5.6 d	4.4 c	32.8 d	12.0 c	8.9 b
Veltyma 3.34 S 7.0 fl oz	5.3 d	4.4 d	4.6 c	39.3 cd	9.8 c	13.1 b
Headline 2.08 SC 6.0 fl oz	8.3 d	9.1 cd	6.6 c	42.8 bcd	17.5 c	12.0 b
Headline AMP 1.68 SE 10.0 fl oz	7.4 d	7.0 cd	5.4 c	40.3 bcd	13.6 c	9.4 b
Aproach Prima 2.34 SC 6.8 fl oz	18.8 c	10.8 c	6.9 c	54.8 b	15.0 c	12.5 b
Miravis Neo 2.5 SE 13.7 fl oz	14.0 c	8.0 cd	6.3 c	48.0 bc	14.1 c	11.3 b
Delaro 325 SC 8.0 fl oz	6.1 d	6.5 cd	5.2 c	36.5 cd	14.8 c	9.8 b
Lucento 4.17 SC 5.0 fl oz	28.0 b	20.1 b	9.7 b	83.8 a	41.5 b	16.8 b
Tilt 3.6 EC 4.0 fl oz	29.0 b	21.2 b	11.5 b	88.5 a	41.8 b	21.4 b
P-value[w]	<.0001	<.0001	<.0001	<.0001	<.0001	<.0001

[z] Fungicide treatments were applied on August 7 at the tassel/silk (VT/R1) growth stage, and all treatments contained a nonionic surfactant (Preference) at a rate of 0.25% v/v.

[y] Tar spot stromata was visually assessed as a percentage (0–100%) of leaf area on five plants in each plot at the ear leaf (EL), ear leaf minus two (EL–2), and ear leaf plus two (EL+2) on October 13.

[x] Tar spot chlorosis and necrosis symptoms were visually assessed as a percentage (0–100%) of leaf area on five plants in each plot at the ear leaf (EL), ear leaf minus two (EL–2), and ear leaf plus two (EL+2) on October 13.

[w] Means followed by the same letter are not significantly different based on Fisher's Least Significant Difference test (LSD; α=0.05).

TABLE 21. *Effect of Fungicide on Canopy Greenness and Corn Yield*

TREATMENT AND RATE/ACRE[z]	CANOPY GREEN (R5)[y] %	CANOPY GREEN (R6)[y] %	MOISTURE %	TEST WEIGHT LB/BU	YIELD[x] BU/ACRE
Nontreated control	85.0	17.5 e	22.9	51.4	203.9
Revytek 3.33 LC 8.0 fl oz	88.8	52.5 ab	24.2	50.5	239.4
Veltyma 3.34 S 7.0 fl oz	90.0	53.8 a	24.3	50.6	240.4
Headline 2.08 SC 6.0 fl oz	90.0	43.8 cd	24.2	50.7	230.7
Headline AMP 1.68 SE 10.0 fl oz	90.0	45.0 bcd	24.2	50.7	223.1
Aproach Prima 2.34 SC 6.8 fl oz	90.0	47.5 abc	24.4	50.9	233.8
Miravis Neo 2.5 SE 13.7 fl oz	90.0	52.5 ab	24.7	50.5	228.0
Delaro 325 SC 8.0 fl oz	91.3	48.8 abc	24.7	50.6	238.1
Lucento 4.17 SC 5.0 fl oz	90.0	41.3 cd	24.0	51.1	231.6
Tilt 3.6 EC 4.0 fl oz	90.0	37.5 d	24.2	51.0	221.8
P-value[w]	0.2047	<.0001	0.0119	0.3853	0.0626

[z] Fungicide treatments were applied on August 7 at the tassel/silk (VT/R1) growth stage, and all treatments contained a nonionic surfactant (Preference) at a rate of 0.25% v/v.

[y] Canopy green was visually assessed as the percentage (0–100%) in the plot on September 22 at dent (R5) and October 13 at full maturity (R6) growth stages.

[x] Yields were adjusted to 14.5% moisture at harvest and adjusted to account for wind damage that caused stand loss in plots where it occurred.

[w] Means followed by the same letter are not significantly different based on Fisher's Least Significant Difference test (LSD; α=0.05).

UNIFORM FUNGICIDE TIMING AND TAR SPOT MODEL VALIDATION IN CORN IN NORTHWESTERN INDIANA, 2020 (COR20-05.PPAC)

T. J. Ross, J. D. Ravellette, S. Shim, and D. E. P. Telenko, Department of Botany and Plant Pathology, Purdue University West Lafayette, Indiana 47907-2054

CORN (*ZEA MAYS* W2585SSRIB)

Tar spot, *Phyllachora maydis*

A trial was established at the Pinney Purdue Agricultural Center (PPAC) in Porter County, Indiana. The experiment was a randomized complete block design with four replications. Plots were 10 feet wide and 30 feet long and consisted of four rows, and the two center rows were used for evaluation. The previous crop was corn. Standard practices for nonirrigated grain corn production in Indiana were followed. Corn hybrid W2585S-SRIB was planted in 30-inch row spacing at a rate of 34,000 seeds/acre on June 8. All fungicide applications were applied at 15 gal/acre and 40 psi using a Lee self-propelled sprayer equipped with a 10-foot boom, fitted with six TJ-VS 8002 nozzles spaced 20 inches apart, at 3.6 mph. Fungicides were applied on July 14, July 20, August 7, August 21, September 2, September 11, and September 23 at the eight-leaf (V8), 10-leaf (V10), tassel/silk (VT/R1), blister (R2), milk (R3), dough (R4), dent (R5), and V8 followed by (fb) VT growth stages, respectively. A prediction model-based treatment was included in the trial, but the model never triggered a fungicide application during the season; therefore, this treatment provided an additional nontreated control for comparison. Disease ratings were assessed on September 22 at the dent (R5) and October 6 at the maturity (R6) growth stages. Tar spot was rated by visually assessing the percentage of stromata and the percentage of symptomatic tissues (chlorosis and necrosis) per leaf on five plants in each at the ear leaf (EL), ear leaf minus 2 leaves (EL–2), and ear leaf plus 2 leaves (EL+2). The values of the five leaves for each plot were averaged before analysis. The two center rows of each plot were harvested on November 4, and yields were adjusted to 15.5% moisture. Data were subjected to mixed model analysis of variance (SAS 9.4, 2019), and means were compared using Fisher's Least Significant Difference test (LSD; α=0.05).

In 2020, weather conditions were favorable for disease at the end of the season but did not trigger the tar spot model application. Tar spot was first detected in the trial on August 4 and was the most prominent disease in the trial, reaching moderate to high severity. Trivapro applied at the V8 fb VT, R2, VT/R1, and R3 growth stages significantly reduced tar spot stromata over the nontreated control on the EL–2 and EL+2 on September 22 (Table 22). Chlorotic and necrotic symptoms of tar spot on the EL–2 and EL+2 on September 22 were significantly reduced by Trivapro applications made at VT/R1, R2, and V8 fb VT growth stages, but no significant differences were detected on the EL. By October 6, Trivapro application at the R2 growth stage had significantly reduced the percentage of stromata on all leaves when compared with other treatments and the nontreated control (Table 23). On October 6, significantly less chlorotic and necrotic symptoms were detected on all leaves for Trivapro applications made at VT/R1, R2, R3, and V8 fb VT growth stages. Trivapro applied at the V8 fb VT, V1/R1, and R2 growth stages significantly increased the percent of canopy greenness of the corn over the nontreated control on both September 22 and October 6 (Table 24). No differences in yield were detected between treatments.

TABLE 22. *Effect of Fungicide on Tar Spot on September 22*

TREATMENT, RATE/ACRE, AND TIMING	TAR SPOT % STROMA[Y] EL-2	TAR SPOT % STROMA[Y] EL	TAR SPOT % STROMA[Y] EL+2	TAR SPOT % CHLOR/ NEC[X] EL-2	TAR SPOT % CHLOR/ NEC[X] EL	TAR SPOT % CHLOR/ NEC[X] EL+2
Nontreated control	26.0 ab	14.8 a	7.5 abc	14.2 bc	7.0	2.2 a
Trivapro 13.7 fl oz at V8	25.5 ab	15.5 a	8.3 abc	17.3 abc	4.6	2.0 ab
Trivapro 13.7 fl oz at V10	22.0 bc	13.5 a	8.8 ab	14.6 abc	5.1	1.1 bcd
Trivapro 13.7 fl oz at VT/R1	11.0 d	6.6 b	5.7 bcd	1.4 d	0.9	0.3 d
Trivapro 13.7 fl oz at R2	10.5 d	5.9 b	3.6 d	1.5 d	0.2	0.2 d
Trivapro 13.7 fl oz at R3	19.0 c	8.0 b	5.1 cd	8.8 cd	1.9	0.7 cd
Trivapro 13.7 fl oz at R4	26.8 a	15.8 a	8.3 abc	21.6 ab	7.5	2.2 a
Trivapro 13.7 fl oz at R5	28.8 a	17.5 a	8.8 ab	15.5 abc	4.6	1.7 abc
Trivapro 13.7 fl oz at V8 fb VT	5.6 e	4.9 b	6.1 bcd	0.7 d	0.4	0.3 d
Trivapro 13.7 fl oz at Model/NTC	28.5a	15.3 a	9.5 a	26.2 a	9.4	1.4 abc
P-value[w]	*<.0001*	*<.0001*	*0.0150*	*0.0009*	*0.1074*	*0.0008*

[z] Fungicide treatments were applied on July 14, July 20, August 7, August 21, September 2, September 11 and September 23 at the V8, V10, tassel/silk (VT/R1), blister (R2), milk (R3), dough (R4), and dent (R5) growth stages, respectively, and all treatments contained a nonionic surfactant (Preference) at a rate of 0.25% v/v. Model/NTC = tar spot weather-based model application. The tar spot model did not cross the action threshold in Indiana during the season; therefore, no fungicide application to this treatment. fb = followed by.

[y] Tar spot stromata was visually assessed as a percentage (0–100%) of leaf area on five plants in each plot at the ear leaf (EL), ear leaf minus two (EL-2), and ear leaf plus two (EL+2).

[x] Tar spot chlorosis and necrosis symptoms were visually assessed as a percentage (0–100%) of leaf area on five plants in each plot at the ear leaf (EL), ear leaf minus two (EL-2), and ear leaf plus two (EL+2).

[w] Means followed by the same letter are not significantly different based on Fisher's Least Significant Difference test (LSD; α=0.05).

TABLE 23. *Effect of Fungicide on Tar Spot on October 6*

TREATMENT, RATE/ACRE, AND TIMING	TAR SPOT % STROMA[y] EL-2	TAR SPOT % STROMA[y] EL	TAR SPOT % STROMA[y] EL+2	TAR SPOT % CHLOR/ NEC[x] EL-2	TAR SPOT % CHLOR/ NEC[x] EL	TAR SPOT % CHLOR/ NEC[x] EL+2
Nontreated control	32.8 ab	29.5 ab	25.3 ab	81.0 a	56.3 a	30.5 a
Trivapro 13.7 fl oz at V8	31.8 abc	30.5 ab	24.0 ab	74.0 ab	53.5 ab	17.5 bc
Trivapro 13.7 fl oz at V10	31.0 abc	29.8 bc	21.5 bc	68.0 ab	53.8 ab	23.5 ab
Trivapro 13.7 fl oz at VT/R1	25.5 d	23.0 c	19.0 cd	29.6 cd	14.5 c	7.8 cd
Trivapro 13.7 fl oz at R2	13.8 g	11.8 e	9.0 e	15.4 d	7.8 c	3.2 d
Trivapro 13.7 fl oz at R3	25.5 d	23.5 c	16.3 d	39.3 c	21.3 c	6.9 cd
Trivapro 13.7 fl oz at R4	28.5 cd	27.5 b	19.3 cd	61.8 b	39.5 b	12.3 cd
Trivapro 13.7 fl oz at R5	29.5 bc	29.3 ab	21.5 bc	74.3 ab	57.0 a	26.5 ab
Trivapro 13.7 fl oz at V8 fb VT	21.8 e	18.5 d	15.8 d	20.0 d	10.3 c	5.5 d
Trivapro 13.7 fl oz at Model/NTC	33.8 a	31.3 a	26.5 a	73.0 ab	53.5 ab	26.8 ab
P-value[w]	<.0001	<.0001	<.0001	<.0001	<.0001	<.0001

[z] Fungicide treatments were applied on July 14, July 20, August 7, August 21, September 2, September 11, and September 23 at the V8, V10, tassel/silk (VT/R1), blister (R2), milk (R3), dough (R4), and dent (R5) growth stages, respectively, and all treatments contained a nonionic surfactant (Preference) at a rate of 0.25% v/v. Model/NTC = tar spot weather-based model application. The tar spot model did not cross the action threshold in Indiana during the season; therefore, no fungicide application to this treatment. fb = followed by.

[y] Tar spot stromata was visually assessed as a percentage (0–100%) of leaf area on five plants in each plot at the ear leaf (EL), ear leaf minus two (EL-2), and ear leaf plus two (EL+2).

[x] Tar spot chlorosis and necrosis symptoms were visually assessed as a percentage (0–100%) of leaf area on five plants in each plot at the ear leaf (EL), ear leaf minus two (EL-2), and ear leaf plus two (EL+2).

[w] Means followed by the same letter are not significantly different based on Fisher's Least Significant Difference test (LSD; α=0.05).

TABLE 24. *Effect of Fungicide on Canopy Greenness and Corn Yield*

TREATMENT, RATE/ACRE, AND TIMING[z]	CANOPY GREEN[y] % SEPTEMBER 22	CANOPY GREEN[y] % OCTOBER 6	MOISTURE %	TEST WEIGHT LB/BU	YIELD[x] BU/ ACRE
Nontreated control	75.0 bcd	12.5 f	22.4	53.3	181.8
Trivapro 13.7 fl oz at V8	80.0 a-d	27.5 e	23.1	52.8	179.0
Trivapro 13.7 fl oz at V10	85.0 abc	27.5 e	23.0	53.5	188.0
Trivapro 13.7 fl oz at VT/R1	90.0 a	42.5 a	24.7	52.9	193.3
Trivapro 13.7 fl oz at R2	87.5 ab	42.5 a	24.3	52.9	193.9
Trivapro 13.7 fl oz at R3	83.8 abc	35.0 bcd	24.5	53.4	185.4
Trivapro 13.7 fl oz at R4	73.8 cd	37.5 abc	23.9	53.2	185.6
Trivapro 13.7 fl oz at R5	70.0 d	28.8 de	22.7	53.3	183.7
Trivapro 13.7 fl oz at V8 fb VT	90.0 a	41.3 ab	23.9	53.2	191.0
Trivapro 13.7 fl oz at Model	75.0 bcd	31.3 cde	22.8	53.0	179.5
P-value[w]	0.0182	<.0001	0.0587	0.7079	0.5435

[z] Fungicide treatments were applied on July 14, July 20, August 7, August 21, September 2, September 11, and September 23 at the V8, V10, tassel/silk (VT/R1), blister (R2), milk (R3), dough (R4), and dent (R5) growth stages, respectively, and all treatments contained a nonionic surfactant (Preference) at a rate of 0.25% v/v. Model/NTC = tar spot weather-based model application. The tar spot model did not cross the action threshold in Indiana during the season; therefore, no fungicide application to this treatment. fb = followed by.

[y] Canopy green was visually assessed as a percentage (0–100%) of crop canopy on September 22 and October 6.

[x] Yields were adjusted to 15.5% moisture at harvest on November 4.

[w] Means followed by the same letter are not significantly different based on Fisher's Least Significant Difference test (LSD; α=0.05).

FUNGICIDE COMPARISON FOR FOLIAR DISEASES IN CORN IN NORTHWESTERN INDIANA, 2020 (COR20-14.PPAC)

S. Shim, J. D. Ravellette, and D. E. P. Telenko, Department of Botany and Plant Pathology, Purdue University West Lafayette, Indiana 47907-2054

CORN (*ZEA MAYS* W2585SSRIB)

Tar spot, *Phyllachora maydis*

A trial was established at the Pinney Purdue Agricultural Center (PPAC) in Porter County, Indiana. The experiment was a randomized complete block design with four replications. Plots were 10 feet wide and 30 feet long and consisted of four rows, and the two center rows were used for evaluation. The previous crop was corn. Standard practices for grain corn production in Indiana were followed. Corn hybrid W2585SSRIB was planted in 30-inch row spacing at a rate of 34,000 seeds/acre on June 8. Standard practices for nonirrigated grain corn production in Indiana were followed. All fungicide applications were applied at 15 gal/acre and 40 psi using a Lee self-propelled sprayer equipped with a 10-foot boom, fitted with six TJ-VS 8002 nozzles spaced 20 inches apart, at 3.6 mph. Fungicides were applied on July 8 at V7, July 28 at V13, and August 8 at the tassel/silk (VT/R1), August 21 at the milk (R2), and September 2 at the milk (R3) growth stages. Disease ratings were assessed on September 17 and September 29 at the dent (R5) and maturity (R6) growth stages, respectively. Tar spot was rated by visually assessing the percentage of stromata and the percentage of symptomatic tissues (chlorosis and necrosis) per leaf on five plants in each plot at the ear leaf (EL), ear leaf minus two (EL–2), and ear leaf plus two (EL+2). Values for each plot were averaged before analysis. The two center rows of each plot were harvested on November 4, and yields were adjusted to 15.5% moisture. Data were subjected to mixed model analysis of variance (SAS 9.4, 2019), and means were compared using Fisher's Least Significant Difference test (LSD; α=0.05).

Tar spot was the most prominent diseases in the trial and reached moderate severity. All fungicides significantly reduced tar spot stromata severity on the EL–2, EL, and EL+2 over the nontreated except Fortix NXT at 6.0 fl oz and Fortix 3.22 SC at 5.0 fl oz applied at V7 on September 17 (Table 25). All fungicides reduced chlorosis and necrosis on EL–2 and EL except Fortix NXT at 6.0 fl oz and Fortix 3.22 SC at 5.0 fl oz applied at V7. No chlorosis or necrosis was noted on EL+2 on September 17. Tar spot stromata severity on all leaves on September 29 was significantly reduced over the nontreated control by all fungicide programs except for Fortix NXT at 6.0 fl oz and Fortix 3.22 SC at 5.0 fl oz applied at V7 (Table 26). All fungicides reduced chlorosis and necrosis on the EL–2 and EL except for the Fortix applications at V7. On the EL+2, Miravis Neo applied at VT/R1, R2, and R3; Trivapro applied at R2; Aproach Prima applied at VT/R1; Aproach applied at V7 followed by Aproach Prima at VT/R1; Dexter Xcel applied at VT/R1; and Headline AMP applied at VT/R1 reduced chlorosis and necrosis over the nontreated on September 29. All fungicide treatments significantly increased the percent of canopy greenness over the nontreated controls on both September 29 and October 7 except for Fortix 3.22 SC at V7 and Miravis Neo at V13 on October 7. There were no significant differences between treatments for lodging and test weight (Table 27). Miravis Neo applied at R2, Trivapro applied at VT/R1, Dexter Xcel applied at VT/R1, and Headline APM applied at VT/R1 significantly increased yield over the nontreated controls.

TABLE 25. *Effect of Fungicide on Tar Spot on September 17*

TREATMENT, RATE/ACRE, AND TIMING[z]	TAR SPOT % STROMA[y] EL−2	TAR SPOT % STROMA[y] EL	TAR SPOT % STROMA[y] EL+2	TAR SPOT % CHLO/ NECR[x] EL−2	TAR SPOT % CHLO/ NECR[x] EL
Nontreated control	10.1 a	6.0 a	2.3 b	2.4 ab	1.2 a
Miravis Neo 2.5 SC 13.7 fl oz at V13	2.7 bcd	1.8 bc	1.6 b	0.1 c	0.0 c
Miravis Neo 2.5 SC 13.7 fl oz at VT/R1	1.1 cd	0.7 c	0.4 c	0.0 c	0.0 c
Miravis Neo 2.5 SC 13.7 fl oz at R2	2.6 cd	1.0 c	0.3 c	0.3 c	0.0 c
Miravis Neo 2.5 SC 13.7 fl oz at R3	5.4 b	2.7 b	1.9 b	1.0 bc	0.2 bc
Trivapro 2.21 SE 13.7 fl oz at VT/R1	0.9 cd	0.5 c	0.4 c	0.0 c	0.0 c
Trivapro 2.21 SE 13.7 fl oz at R2	2.8 bcd	0.7 c	0.2 c	0.5 bc	0.0 c
Aproach Prima 2.34 SC 6.8 fl oz at VT/R1	0.6 cd	0.6 c	0.4 c	0.0 c	0.0 c
Aproach 2.08 SC 6.0 fl oz at V8 fb Aproach Prima 2.34 SC 6.8 fl oz at VT/R1	0.6 d	0.3 c	0.2 c	0.1c	0.0 c
Fortix NXT 6.0 fl oz at VT/R1	1.1 cd	0.5 c	0.2 c	0.0 c	0.0 c
Zolera ODX 5.0 fl oz at VT/R1	1.5 cd	0.5 c	0.2 c	0.2 c	0.0 c
Dexter Xcel 48.0 fl oz at VT/R1	0.9 cd	0.6 c	0.2 c	0.0 c	0.0 c
Zolera FX 5.0 fl oz at VT/R1	1.3 cd	0.6 c	0.4 c	0.0 c	0.0 c
Fortix NXT 6.0 fl oz at V7	10.3 a	5.8 a	3.0 a	4.3 a	0.1 bc
Fortix 3.22 SC 5.0 fl oz at V7	10.1 a	6.2 a	3.2 a	3.9 a	0.8 ab
Headline AMP 1.68 SC 10.0 fl oz at VT/R1	3.4 bc	1.5 bc	0.6 c	0.0 c	0.0 c
Nontreated control	10.9 a	7.2 a	3.1 a	3.6 a	1.2 a
P-value[w]	<.0001	<.0001	<.0001	<.0001	0.0163

[z] Fungicide treatments were applied on July 8 at V7, July 28 at V13, August 8 at tassel/silk (VT/R1), August 21 at blister (R2), and September 2 at milk (R3) growth stages, and all treatments contained a nonionic surfactant (Preference) at a rate of 0.25% v/v. fb = followed by.

[y] Tar spot stromata was visually assessed as a percentage (0–100%) of leaf area on five plants in each plot at the ear leaf (EL), ear leaf minus two (EL−2), and ear leaf plus two (EL+2).

[x] Tar spot chlorosis and necrosis symptoms were visually assessed as a percentage (0–100%) of leaf area on five plants in each plot at the ear leaf (EL), ear leaf minus two (EL−2), and ear leaf plus two (EL+2).

[w] Means followed by the same letter are not significantly different based on Fisher's Least Significant Difference test (LSD; α=0.05).

TABLE 26. *Effect of Fungicide on Tar Spot on September 29*

TREATMENT, RATE/ACRE, AND TIMING[z]	TAR SPOT % STROMA[y] EL–2	TAR SPOT % STROMA[y] EL	TAR SPOT % STROMA[y] EL+2	TAR SPOT % CHLO/ NECR[x] EL–2	TAR SPOT % CHLO/ NECR X EL	TAR SPOT % CHLO/ NECR X EL+2
Nontreated control	29.0 a	18.9 a	7.7 a	77.3 a	36.8 a	14.3 a
Miravis Neo 2.5 SC 13.7 fl oz at V13	12.0 c	6.3 c	4.0 cd	40.0 b	13.6 b	5.8 bcd
Miravis Neo 2.5 SC 13.7 fl oz at VT/R1	4.0 d	3.1 cd	2.1 de	24.8 bcd	11.2 b	4.5 cd
Miravis Neo 2.5 SC 13.7 fl oz at R2	4.3 d	1.9 d	1.0 ef	21.8 bcd	9.2 b	4.0 d
Miravis Neo 2.5 SC 13.7 fl oz at R3	12.2 c	4.4 cd	1.9 ef	37.4 bc	11.9 b	4.7 cd
Trivapro 2.21 SE 13.7 fl oz at VT/R1	3.9 d	3.4 cd	2.1 ef	22.3 bcd	7.4 b	7.1 bcd
Trivapro 2.21 SE 13.7 fl oz at R2	3.0 d	1.6 d	0.3 f	31.5 bcd	7.0 b	3.9 d
Aproach Prima 2.34 SC 6.8 fl oz at VT/R1	3.1 d	2.2 d	1.5 ef	22.1 bcd	8.9 b	5.2 cd
Aproach 2.08 SC 6.0 fl oz at V8 fb Aproach Prima 2.34 SC 6.8 fl oz at VT/R1	3.5 d	2.3 d	1.8 ef	25.4 bcd	9.2 b	2.9 d
Fortix NXT 6.0 fl oz at VT/R1	3.4 d	1.0 d	0.9 ef	16.6 d	5.8 b	5.9 bcd
Zolera ODX 5.0 fl oz at VT/R1	2.7 d	1.3 d	0.5 ef	18.4 cd	5.6 b	3.7 d
Dexter Xcel 48.0 fl oz at VT/R1	2.4 d	1.7 d	0.8 ef	25.1 bcd	5.7 b	2.2 d
Zolera FX 5.0 fl oz at VT/R1	1.7 d	1.3 d	1.2 ef	25.6 bcd	10.5 b	6.8 bcd
Fortix NXT 6.0 fl oz at V7	22.4 b	20.0 a	7.3 ab	60.8 a	30.8 a	9.6 abc
Fortix 3.22 SC 5.0 fl oz at V7	23.9 b	14.6 b	5.7 bc	68.3 a	28.4 a	9.8 abc
Headline AMP 1.68 SC 10.0 fl oz at VT/R1	4.8 d	2.7 d	2.3 de	31.6 bcd	9.3 b	4.4 cd
Nontreated control	31.3 a	21.3 a	7.3 ab	65.5 a	32.4 a	10.6 ab
P-value[w]	<.0001	<.0001	<.0001	<.0001	<.0001	0.0037

[z] Fungicide treatments were applied on July 8 at V7, July 28 at V13, August 8 at tassel/silk (VT/R1), August 21 at blister (R2), and September 2 at milk (R3) growth stages, and all treatments contained a nonionic surfactant (Preference) at a rate of 0.25% v/v. fb = followed by.

[y] Tar spot stromata was visually assessed as a percentage (0–100%) of leaf area on five plants in each plot at the ear leaf (EL), ear leaf minus two (EL–2), and ear leaf plus two (EL+2). fb = followed by.

[x] Tar spot chlorosis and necrosis symptoms were visually assessed as a percentage (0–100%) of leaf area on five plants in each plot at the ear leaf (EL), ear leaf minus two (EL–2), and ear leaf plus two (EL+2).

[w] Means followed by the same letter are not significantly different based on Fisher's Least Significant Difference test (LSD; α=0.05).

TABLE 27. *Effect of Fungicide on Canopy Greenness, Lodging, and Corn Yield*

TREATMENT, RATE/ACRE, AND TIMING[z]	CANOPY GREEN (R5)[y] %	CANOPY GREEN (R6)[y] %	LODGING[x] %	HARVEST MOISTURE %	TEST WEIGHT LB/BU	YIELD[w] BU/ACRE
Nontreated control	60.0 e	26.3 de	5.0	22.1 f	52.5	179.5 c-f
Miravis Neo 2.5 SC 13.7 fl oz at V13	72.5 bc	31.3 bcd	0.0	23.4 c-f	52.7	193.9 abc
Miravis Neo 2.5 SC 13.7 fl oz at VT/R1	78.8 ab	38.8 ab	0.0	25.3 ab	52.9	187.0 bcd
Miravis Neo 2.5 SC 13.7 fl oz at R2	76.3 ab	41.3 a	0.0	25.3 ab	52.0	197.5 ab
Miravis Neo 2.5 SC 13.7 fl oz at R3	75.0 ab	40.0 a	2.5	24.1 a-e	52.2	179.8 c-f
Trivapro 2.21 SE 13.7 fl oz at VT/R1	78.8 ab	37.5 ab	0.0	24.5 a-d	52.5	200.0 ab
Trivapro 2.21 SE 13.7 fl oz at R2	78.8 ab	41.3 a	5.0	24.6 abc	52.6	181.7 cde
Aproach Prima 2.34 SC 6.8 fl oz at VT/R1	78.8 ab	42.5 a	0.0	25.5 a	52.4	185.3 c-e
Aproach 2.08 SC 6.0 fl oz at V8 fb Aproach Prima 2.34 SC 6.8 fl oz at VT/R1	80.0 a	40.0 a	0.0	25.0 ab	51.5	192.7 abc
Fortix NXT 6.0 fl oz at VT/R1	78.8 ab	41.3 a	0.0	24.6 abc	52.3	181.4 cde
Zolera ODX 5.0 fl oz at VT/R1	78.8 ab	38.8 ab	0.0	24.5 a-d	53.1	192.1 abc
Dexter Xcel 48.0 fl oz at VT/R1	78.8 ab	41.3 a	2.5	24.4 a-d	52.3	202.2 a
Zolera FX 5.0 fl oz at VT/R1	78.8 ab	36.3 abc	2.5	24.1 b-e	53.3	176.6 def
Fortix NXT 6.0 fl oz at V7	67.5 cd	28.8 cd	0.0	22.9 ef	52.8	180.7 c-f
Fortix 3.22 SC 5.0 fl oz at V7	61.3 de	20.0 e	2.5	23.2 def	53.3	171.4 ef
Headline AMP 1.68 SC 10.0 fl oz at VT/R1	77.5 ab	42.5 a	2.5	24.4 a-d	52.6	199.8 ab
Nontreated control	57.5 e	23.8 de	10.0	22.6 f	52.4	165.8 f
P-value[v]	<.0001	<.0001	0.4486	<.0002	0.4875	0.0001

[z] Fungicide treatments were applied on July 8 at V7, July 28 at V13, August 8 at tassel/silk (VT/R1), August 21 at blister (R2) and September 2 at milk (R3) growth stages, and all treatments contained a nonionic surfactant (Preference) at a rate of 0.25% v/v. fb = followed by.

[y] Canopy green was visually assessed as a percentage (0–100%) of crop canopy on September 29 at dent (R5) and October 7 at maturity (R6) growth stages.

[x] Lodging = percentage of lodged stalks when pushed from shoulder height to 45° from vertical on September 29.

[w] Yields were adjusted to 15.5% moisture at harvest on November 4.

[v] Means followed by the same letter are not significantly different based on Fisher's Least Significant Difference test (LSD; α=0.05).

FUNGICIDE EFFICACY AND TIMING FOR TAR SPOT OF CORN IN NORTHWESTERN INDIANA, 2020 (COR20-15.PPAC)

C. Rocca Da Silva, J. D. Ravellette, S. Shim, and D. E. P. Telenko, Department of Botany and Plant Pathology, Purdue University West Lafayette, Indiana 47907-2054

CORN (*ZEA MAYS* W2585SSRIB)

Tar spot, *Phyllachora maydis*

A trial was established at the Pinney Purdue Agricultural Center (PPAC) in Porter County, Indiana. The experiment was a randomized complete block design with four replications. Plots were 10 feet wide and 30 feet long and consisted of four rows, and the two center rows were used for evaluation. The previous crop was corn. Standard practices for grain corn production in Indiana were followed. Corn hybrid W2585SSRIB was planted in 30-inch row spacing at a rate of 34,000 seeds/acre on June 6 using a GPS-guided John Deere 1700 six-row planter. The field was overhead irrigated weekly at 1 inch unless weekly rainfall was 1 inch or higher to encourage disease. All fungicide applications were applied at 15 gal/acre and 40 psi using a Lee self-propelled sprayer equipped with a 10-foot boom, fitted with six TJ-VS 8002 nozzles spaced 20 inches apart, at 3.6 mph. Fungicides were applied on July 14 at V8, August 5 at V8 plus 3 weeks after treatment (3-WAT), August 5 at first detection of tar spot, August 7 at tassel/silk (VT/R1), August 27 at first detection plus 3-WAT, August 27 at VT plus 3-WAT, September 2 at milk (R3), and September 23 at R3 plus 3-WAT. Disease ratings were assessed on October 6 at maturity (R6) growth stage. Tar spot was rated by visually assessing the percentage of stromata and the percentage of symptomatic tissues (chlorosis and necrosis) per leaf on five plants in each plot at the ear leaf. Values for each plot were averaged before analysis. The two center rows of each plot were harvested on November 6, and yields were adjusted to 15.5% moisture. Data were subjected to mixed model analysis of variance (SAS 9.4, 2019), and means were compared using Fisher's Least Significant Difference test (LSD; α=0.05).

In 2020, weather conditions were favorable for disease. Tar spot was first detected in the trial on July 28 and was the most prominent disease in the trial, reaching moderate to high severity. Veltyma and Lucento applied at the first detection of tar spot, VT and V8 followed by (fb) 3-WAT, significantly reduced tar spot stromata over the nontreated control on the ear leaf on October 6 (Table 28). Chlorotic and necrotic symptoms of tar spot on the ear leaf on October 6 were significantly reduced by Veltyma applications applied at the first detection of tar spot, VT, and the first detection fb 3-WAT, V8 fb 3-WAT, and VT fb 3-WAT, and by Lucento when applied at the first detection of tar spot, VT, R3, and the first detection fb 3-WAT, V8 fb 3-WAT, and VT fb 3-WAT. Veltyma and Lucento significantly increased the percentage of green canopy over the nontreated control on October 6 except when applied at V8. No difference between treatments and nontreated control were detected for harvest moisture, test weight, and corn yield.

TABLE 28. *Effect of Fungicide on Tar Spot, Canopy Greenness, and Corn Yield*

TREATMENT, RATE/ACRE, AND TIMING[z]	TAR SPOT % SEVER-ITY[y]	TAR SPOT % CHLOR/NEC[x]	CANOPY GREEN[w] %	HARVEST MOISTURE %	TEST WEIGHT LB/BU	YIELD LB/ACRE
Nontreated control	31.5 ab	52.3 ab	23.8 i	23.9	51.5	220.5
Veltyma 3.34 S 7.0 fl oz at first detection	9.0 g	1.3 g	80.0 a	24.5	50.5	224.8
Veltyma 3.34 S 7.0 fl oz at V8	31.3 ab	41.8 bc	30.0 hi	24.0	51.1	219.4
Veltyma 3.34 S 7.0 fl oz at VT	8.3 g	0.7 g	76.3 ab	23.9	50.8	219.4
Veltyma 3.34 S 7.0 fl oz at R3	27.8 bc	23.4 de	43.8 efg	23.9	51.1	218.9
Veltyma 3.34 S 7.0 fl oz at first detection fb 3-WAT	16.5 f	4.6 g	65.0 bc	24.2	50.8	225.7
Veltyma 3.34 S 7.0 fl oz at V8 fb 3-WAT	9.5 g	1.0 g	77.5 a	24.0	50.5	228.7
Veltyma 3.34 S 7.0 fl oz at VT fb 3-WAT	19.5 ef	5.9 fg	55.0 cde	23.7	50.9	223.5
Veltyma 3.34 S 7.0 fl oz at R3 fb 3-WAT	31.3 ab	34.0 cd	28.8 hi	23.4	50.7	221.0
Nontreated control	33.0 a	61.5 a	22.5 i	22.5	51.4	220.1
Lucento 7.17 SC 5.0 fl oz at first detection	21.8 de	9.5 efg	56.3 cd	24.5	50.7	225.5
Lucento 7.17 SC 5.0 fl oz at V8	30.5 ab	29.8 cd	31.3 hi	23.3	51.4	213.2
Lucento 7.17 SC 5.0 fl oz at VT	20.8 ef	6.4 efg	48.8 def	23.2	51.2	225.6
Lucento 7.17 SC 5.0 fl oz at R3	20.0 ef	9.3 efg	43.8 efg	23.6	51.3	218.1
Lucento 7.17 SC 5.0 fl oz at first detection fb 3-WAT	23.3 de	7.8 ef	53.8 cde	24.4	51.7	224.0
Lucento 7.17 SC 5.0 fl oz at V8 fb 3-WAT	23.3 de	11.4 efg	53.8 cde	23.7	50.7	233.5
Lucento 7.17 SC 5.0 fl oz at VT fb 3-WAT	25.5 cd	23.1 def	38.8 fgh	24.2	51.1	219.8
Lucento 7.17 SC 5.0 fl oz at R3 fb 3-WAT	30.3 ab	33.0 cd	33.8 ghi	23.9	50.9	216.1
P-value[u]	<.0001	<.0001	<.0001	0.1621	0.4319	0.8278

[z] Fungicides were applied on July 14 at V8, August 5 at V8 plus 3 weeks after treatment (3-WAT), August 5 at first detection of tar spot, August 7 at tassel/silk (VT/R1), August 27 at first detection plus 3-WAT, August 27 at VT plus 3-WAT, September 2 at milk (R3), and September 23 at R3 plus 3-WAT. All treatments contained a nonionic surfactant (Preference) at a rate of 0.25% v/v. fb = followed by.

[y] Tar spot stromata was visually assessed as a percentage (0–100%) of leaf area on five plants in each plot at the ear leaf on October 6.

[x] Tar spot chlorosis and necrosis symptoms were visually assessed as a percentage (0–100%) of leaf area on five plants in each plot at the ear leaf on October 6.

[w] Canopy green was visually assessed as a percentage (0–100%) of crop canopy on October 6.

[v] Yields were adjusted to account for wind damage that caused stand loss in plots where it occurred and were harvested on November 6.

[u] Means followed by the same letter are not significantly different based on Fisher's Least Significant Difference test (LSD; α=0.05).

EVALUATION OF FUNGICIDES AND APPLICATION TIMING FOR TAR SPOT OF CORN IN NORTHWESTERN INDIANA, 2020 (COR20-23.PPAC)

K. G. Waibel, S. Shim, J. D. Ravellette, and D. E. P. Telenko, Department of Botany and Plant Pathology, Purdue University West Lafayette, Indiana 47907-2054

CORN (*ZEA MAYS* W2585SSRIB)

Tar spot, *Phyllachora maydis*

A trial was established at the Pinney Purdue Agricultural Center (PPAC) in Porter County, Indiana. The trial was a randomized complete block design with four replications. Plots were 10 feet wide and 30 feet long and consisted of four rows, and the two center rows were used for evaluation. The previous crop was corn. Standard practices for grain corn production in Indiana were followed. Corn hybrid W2585SSRIB was planted in 30-inch row spacing at a rate of 34,000 seeds/acre on June 9. The field was overhead irrigated weekly at 1 inch unless weekly rainfall was 1 inch or higher to encourage disease. All fungicide applications were applied at 15 gal/acre and 40 psi using a Lee self-propelled sprayer equipped with a 10-foot boom, fitted with six TJ-VS 8002 nozzles spaced 20 inches apart, at 3.6 mph. Fungicides were applied on July 8 at V7 and on August 8 at tassel/silk (VT/R1) growth stages, respectively. Disease ratings were assessed on September 22 and October 7 at dent (R5) and maturity (R6) growth stages, respectively. Disease severity and chlorosis/necrosis were rated by visually assessing the percentage of symptomatic leaf area on five plants in each plot at the ear leaf (EL), ear leaf minus two (EL–2), and ear leaf plus two (EL+2). Values for each plot were averaged before analysis. Percent lodging was determined from 10 plants in each plot when pushed from shoulder height to 45° from vertical. The two center rows of each plot were harvested on November 6, and yields were adjusted to 15.5% moisture. Data were subjected to mixed model analysis of variance (SAS 9.4, 2019), and means were compared using Fisher's Least Significant Difference test (LSD; α=0.05).

In 2020, tar spot reached moderate severity. All fungicide programs significantly reduced tar spot on October 7 on EL (Table 29). All fungicide programs reduced tar spot chlorotic and necrotic symptoms on October 7. All fungicide program increased percent green over the nontreated control on October 7. Veltyma resulted in the highest percentage of green canopy over other treatments but was only significantly different from Aproach Prima and Trivapro. There were no significant differences between treatments for harvest moisture, test weight, and yield of corn.

TABLE 29. *Effect of Fungicide on Tar Spot, Canopy Greenness, and Corn Yield*

TREATMENT, RATE/ACRE, AND TIMING[z]	TAR SPOT % STROMA[y]	TAR SPOT % CHLOR/ NEC[x]	CANOPY GREEN[w] %	HARVEST MOISTURE %	TEST WEIGHT LB/BU	YIELD[v] BU/ACRE
Nontreated control	25.6 a	44.8 a	47.5 d	24.3	50.6	225.8
Trivapro 2.21 SE 13.7 fl oz at VT/R1	4.8 b	3.9 b	82.5 bc	24.7	50.6	221.7
Aproach Prima 2.34 SC 6.8 fl oz at VT/R1	4.9 b	2.7 b	81.3 c	24.3	50.7	229.1
Aproach 2.08 SC 6.0 fl oz at V7 fb Aproach Prima 2.34 SC 6.8 fl oz at VT/R1	4.8 b	3.3 b	85.0 abc	25.0	50.3	225.8
Delaro Complete 458 SC 8.0 fl oz at VT/R1	5.0 b	6.5 b	87.5 ab	24.9	50.6	240.0
Delaro Complete 458 SC 12.0 fl oz at VT/R1	2.6 b	1.8 b	85.0 abc	24.7	50.2	221.9
Veltyma 3.34 S 7.0 fl oz at VT/R1	2.4 b	0.9 b	88.8 a	24.7	50.4	227.0
Miravis Neo 2.5 SE 13.7 fl oz at VT/R1	4.7 b	1.5 b	83.8 abc	24.6	50.7	225.3
P-value[u]	<.0001	<.0001	<.0001	0.2803	0.4227	0.4019

[z] Fungicide treatments were applied on July 8 at V7 (tassel) and on August 8 at tassel/silk (VT/R1) growth stages and contained a nonionic surfactant (Preference) at a rate of 0.25% v/v. fb = followed by.

[y] Tar spot stromata was visually assessed as a percentage (0–100%) of leaf area on five plants in each plot at the ear leaf on October 7.

[x] Tar spot chlorotic and necrotic symptoms were visually assessed as a percentage (0–100%) of leaf area on five plants in each plot at the ear leaf on October 7.

[w] Canopy green was visually assessed as a percentage (0–100%) of crop canopy on October 7.

[v] Yields were adjusted to 15.5% moisture at harvest on November 6.

[u] Means followed by the same letter are not significantly different based on Fisher's Least Significant Difference test (LSD; α=0.05).

EVALUATION OF FUNGICIDE EFFICACY AND TIMING FOR TAR SPOT OF CORN IN NORTHWESTERN INDIANA, 2020 (COR20-27.PPAC)

D. E. P. Telenko, S. Shim, and J. D. Ravellette, Department of Botany and Plant Pathology, Purdue University West Lafayette, Indiana 47907-2054

CORN (*ZEA MAYS* W2585SSRIB)

Tar spot, *Phyllachora maydis*

A trial was established at the Pinney Purdue Agricultural Center (PPAC) in Porter County, Indiana. The experiment was a randomized complete block design with four replications. Plots were 10 feet wide and 30 feet long and consisted of four rows, and the two center rows were used for evaluation. The previous crop was corn. Standard practices for grain corn production in Indiana were followed. Corn hybrid W2585SSRIB was planted in 30-inch row spacing at a rate of 34,000 seeds/acre on June 9. The field was overhead irrigated weekly at 1 inch unless weekly rainfall was 1 inch or higher to encourage disease. All fungicide applications were applied at 15 gal/acre and 40 psi using a Lee self-propelled sprayer equipped with a 10-foot boom, fitted with six TJ-VS 8002 nozzles spaced 20 inches apart, at 3.6 mph. Fungicides were applied on July 20, August 8, and August 20 at the V10, tassel/silk (VT/R1), and milk (R2) growth stages, respectively. Disease ratings were assessed on October 7 at maturity (R6) growth stage. Tar spot was rated by visually assessing the percentage of stromata and the percentage of symptomatic tissues (chlorosis and necrosis) per leaf on five plants in each plot at the ear leaf. Values for each plot were averaged before analysis. The two center rows of each plot were harvested on November 6, and yields were adjusted to 15.5% moisture. Data were subjected to mixed model analysis of variance (SAS 9.4, 2019), and means were compared using Fisher's Least Significant Difference test (LSD; α=0.05).

In 2020, weather conditions were favorable for disease. Tar spot was the most prominent disease in the trial and reached moderate to high severity. All fungicide programs reduced tar spot stromata on October 7 over the nontreated control (Table 30). Veltyma applied at VT/R1 and R2 and Delaro plus Luna Privilege applied at VT/R1 had the lowest amount of tar spot on the ear leaf but were not significantly different from Miravis Neo applied at VT/R1 and R2 and Delaro plus Luna Privilege applied at R2. All fungicides applied at VT/R1 and R2 significantly reduced tar spot stromata and chlorotic and necrotic symptoms over those fungicides applied at V10 except for Veltyma at V10 on October 7. All fungicide programs increased canopy greenness on October 7 over the nontreated control. Fungicides applied at VT/R1 or R2 were still 70% or more green by October 7. There were no significant differences in test weight. Veltyma and Miravis Neo applied at VT/R1 or R2 resulted in in higher harvest moisture than the nontreated control. Corn yield was highest in plots treated with Miravis Neo and Delaro plus Luna Privilege applied at both VT/R1 and R2, but these were not significantly different from Veltyma applied at VT/R1 or R2.

TABLE 30. *Effect of Fungicide on Tar Spot, Canopy Greenness, and Corn Yield*

TREATMENT, RATE/ACRE, AND TIMING[z]	TAR SPOT % STROMA[y]	TAR SPOT % CHLOR/NEC[x]	CANOPY GREEN[w] %	HARVEST MOISTURE %	TEST WEIGHT LB/BU	YIELD[v] BU/ ACRE
Nontreated control	23.0 a	44.0 a	45.0 d	25.0 c	49.7	201.4 d
Veltyma 7.0 fl oz at V10	6.0 d	5.2 d	60.0 b	25.5 abc	49.5	202.6 cd
Miravis Neo 13.7 fl oz at V10	17.8 b	28.0 b	50.0 cd	25.6 abc	49.5	207.3 bcd
Delaro 8.0 fl oz + Luna Privilege 2.0 fl oz at V10	12.5 c	18.3 c	52.5 c	25.2 bc	49.4	209.3 a-d
Veltyma 7.0 fl oz at VT/R1	1.0 e	0.2 d	76.3 a	25.7 ab	49.3	209.1 a-d
Miravis Neo 13.7 fl oz at VT/R1	4.1 de	1.8 d	70.0 a	26.0 a	49.4	214.5 ab
Delaro 8.0 fl oz + Luna Privilege 2.0 fl oz at VT/R1	1.6 e	0.5 d	76.3 a	25.5 abc	54.6	215.7 ab
Veltyma 7.0 fl oz at R2	1.6 e	0.7 d	71.3 a	25.9 a	49.2	211.1 a-d
Miravis Neo 13.7 fl oz at R2	3.8 de	2.0 d	75.0 a	25.8 ab	49.5	218.1 a
Delaro 8.0 fl oz + Luna Privilege 2.0 fl oz at R2	2.6 de	1.0 d	75.0 a	25.7 abc	49.4	212.6 abc
P-value[u]	<.0001	<.0001	<.0001	<.0001	0.4527	0.0405

[z] Fungicides were applied on July 20, August 8, and August 20 at the V10, tassel/silk (VT/R1), and milk (R2) growth stages, respectively. All fungicide treatments contained BAS 92740S @ 6.4 fl oz/acre.

[y] Tar spot stromata was visually assessed as a percentage (0–100%) of leaf area on five plants in each plot at the ear leaf on October 7.

[x] Tar spot chlorotic and necrotic symptoms were visually assessed as a percentage (0–100%) of leaf area on five plants in each plot at the ear leaf on October 7.

[w] Canopy green was visually assessed as a percentage (0–100%) of crop canopy on October 7.

[v] Yields were adjusted to 15.5% moisture at harvest on November 6.

[u] Means followed by the same letter are not significantly different based on Fisher's Least Significant Difference test (LSD; α=0.05).

EVALUATION OF FUNGICIDE EFFICACY FOR TAR SPOT OF CORN IN NORTHWESTERN INDIANA, 2020 (COR20-28.PPAC)

D. E. P. Telenko, S. Shim, and J. D. Ravellette, Department of Botany and Plant Pathology, Purdue University West Lafayette, Indiana 47907-2054

CORN (*ZEA MAYS* W2585SSRIB)

Tar spot, *Phyllachora maydis*

A trial was established at the Pinney Purdue Agricultural Center (PPAC) in Porter County, Indiana. The experiment was a randomized complete block design with four replications. Plots were 10 feet wide and 30 feet long and consisted of four rows, and the two center rows were used for evaluation. The previous crop was corn. Standard practices for grain corn production in Indiana were followed. Corn hybrid W2585SSRIB was planted in 30-inch row spacing at a rate of 34,000 seeds/acre on June 9. The field was overhead irrigated weekly at 1 inch unless weekly rainfall was 1 inch or higher to encourage disease. All fungicide applications were applied at 15 gal/acre and 40 psi using a Lee self-propelled sprayer equipped with a 10-foot boom, fitted with six TJ-VS 8002 nozzles spaced 20 inches apart, at 3.6 mph. Fungicides were applied on August 8 at tassel/silk (VT/R1) growth stages. Disease ratings were assessed on October 7 at maturity (R6) growth stage. Tar spot was rated by visually assessing the percentage of stromata and the percentage of symptomatic tissues (chlorosis and necrosis) per leaf on five plants in each plot at the ear leaf. Values for each plot were averaged before analysis. The two center rows of each plot were harvested on November 6, and yields were adjusted to 15.5% moisture. Data were subjected to mixed model analysis of variance (SAS 9.4, 2019), and means were compared using Fisher's Least Significant Difference test (LSD; α=0.05).

In 2020, weather conditions were favorable for disease. Tar spot was the most prominent disease in the trial and reached moderate level. All fungicide programs reduced tar spot stromata over the nontreated control (Table 31). Veltyma at 7.0 fl oz had the lowest percent of stromata but was not significantly different from Revytek, Delaro plus Luna Privilege, and Veltyma at 9.0 fl oz. All fungicides reduced chlorotic and necrotic symptoms on all leaves on October 7. Lucento had significantly more symptoms than all other fungicide treatments. All fungicide programs increased canopy greenness on October 7. The Revytek treatment was the greenest but was not significantly different from Veltyma at both 7.0 and 9.0 fl oz, Miravis Neo, and Delaro plus Luna Privilege. All fungicide treatments increased yield over the nontreated control. Corn yield was highest in plots with Revytek, but this was not significantly different from Veltyma at both 7.0 and 9.0 fl oz, Headline AMP, Miravis Neo, and Delaro plus Luna Privilege.

TABLE 31. *Effect of Fungicide on Tar Spot, Canopy Greenness, and Corn Yield*

TREATMENT AND RATE/ACRE[z]	TAR SPOT % STROMA[y]	TAR SPOT % CHLOR/ NEC[x]	CANOPY GREEN[w] %	HARVEST MOISTURE %	TEST WEIGHT LB/BU	YIELD[v] BU/ ACRE
Nontreated control	28.8 a	52.8 a	37.5 d	24.4 c	50.5	200.7 d
Veltyma 3.34 S 7.0 fl oz	1.5 e	0.4 d	81.3 ab	25.3 ab	49.8	230.3 abc
Revytek 3.33 LC 8.0 fl oz	2.4 de	0.5 d	82.5 a	25.3 ab	49.9	237.6 a
Headline AMP 1.68 SC 10.0 fl oz	4.4 cd	1.8 cd	75.0 b	25.1 b	49.6	230.3 abc
Miravis Neo 2.5 SE 13.7 fl oz	5.5 bc	2.5 bc	76.3 ab	25.1 b	49.9	224.7 abc
Delaro 325 SC 8.0 fl oz + Luna Privilege 2.0 fl oz	2.6 de	1.0 cd	78.8 ab	25.3 ab	49.6	233.1 ab
Lucento 4.17 SC 5.0 fl oz	6.7 b	3.7 b	66.3 c	25.1 b	50.2	222.2 bc
Trivapro 2.21 SE 13.7 fl oz	5.0 bc	1.3 cd	75.0 b	25.4 ab	49.6	216.1 c
Veltyma 3.34 S 9.0 fl oz	2.6 de	0.3 d	80.0 ab	25.7 a	49.9	223.5 abc
P-value[u]	<.0001	<.0001	<.0001	0.0155	0.1189	0.0019

[z] Fungicide treatments were applied on August 8 at tassel/silk (VT/R1) growth stage. All fungicide treatments contained a nonionic surfactant (Preference) at a rate of 0.25% v/v.

[y] Tar spot stromata was visually assessed as a percentage (0–100%) of leaf area on five plants in each plot at the ear leaf on October 7.

[x] Tar spot chlorotic and necrotic symptoms were visually assessed as a percentage (0–100%) of leaf area on five plants in each plot at the ear leaf on October 7.

[w] Canopy green was visually assessed as a percentage (0–100%) of crop canopy green on October 7.

[v] Yields were adjusted to 15.5% moisture and to account for wind damage that caused stand loss in plots where it occurred and harvested on November 6.

[u] Means followed by the same letter are not significantly different based on Fisher's Least Significant Difference test (LSD; α=0.05).

ASSESSMENT OF FUNGICIDES APPLIED AT MILK (R3) FOR TAR SPOT OF CORN IN NORTHWESTERN INDIANA, 2020 (COR20-29.PPAC)

K. G. Waibel, J. D. Ravellette, S. Shim, and D. E. P. Telenko, Department of Botany and Plant Pathology, Purdue University West Lafayette, Indiana 47907-2054

CORN (*ZEA MAYS* W2585SSRIB)

Tar spot, *Phyllachora maydis*

A trial was established at the Pinney Purdue Agricultural Center (PPAC) in Porter County, Indiana. The experiment was a randomized complete block design with four replications. Plots were 10 feet wide and 30 feet long and consisted of four rows, and the two center rows were used for evaluation. The previous crop was corn. Standard corn production practices for Indiana were followed. Corn hybrid W2585SSRIB was planted in 30-inch row spacing at a rate of 34,000 seeds/acre on June 8. All fungicide applications were applied at 15 gal/acre and 40 psi using a Lee self-propelled sprayer equipped with a 10-foot boom, fitted with six TJ-VS 8002 nozzles spaced 20 inches apart, at 3.6 mph. Fungicides were applied on September 2 at the milk (R3) growth stage. Disease ratings were assessed on September 29 at the dent (R5) growth stage. Disease severity and chlorosis/necrosis was rated by visually assessing the percentage of symptomatic leaf area on five plants in each plot at the ear leaf. Values for each plot were averaged before analysis. The two center rows of each plot were harvested on November 4, and yields were adjusted to 15.5% moisture. Data were subjected to mixed model analysis of variance (SAS 9.4, 2019), and means were compared using Fisher's Least Significant Difference test (LSD; α=0.05).

In 2020, tar spot reached moderate severity. All fungicides reduced the percent of tar spot stromata over the nontreated control except for Domark plus Badge on September 29 (Table 32). Affiance and Domark treatments resulted in the lowest amount of tar spot over other fungicide treatments on September 29 but were not significantly different from Affiance plus Badge treatment. Percentage of green canopy was significantly increased in all treatments on October 7 over the nontreated control. There was no significant treatment effect on moisture and test weight of corn. Domark was the only treatment that significantly increased yield over the nontreated control.

TABLE 32. *Effect of Fungicide on Tar Spot, Canopy Greenness, and Corn Yield*

TREATMENT AND RATE/ACRE[z]	TAR SPOT % SEVER- ITY[y]	TAR SPOT % CHLOR/ NEC[x]	CANOPY GREEN[w] %	HARVEST MOISTURE %	TEST WEIGHT LB/BU	YIELD[v] BU/ACRE
Nontreated control	21.0 a	31.5	18.8 b	22.0	52.1	170.0 b
Affiance 1.5 SC 10.0 fl oz	9.3 c	13.6	31.3 a	22.8	52.6	177.7 ab
Domark 230 ME 6.0 fl oz	7.7 c	14.0	31.3 a	22.7	52.0	184.6 a
Affiance 1.5 SC 10.0 fl oz + Badge SC 2.0 pt	13.4 bc	19.0	31.3 a	23.0	51.9	176.6 ab
Domark 230 ME 6.0 fl oz + Badge SC 2.0 pt	16.1 ab	19.6	27.5 a	22.2	51.4	173.7 b
P-value[U]	0.0043	0.3438	0.0182	0.5258	0.4677	0.0363

[z] Fungicide treatments were applied on September 2 at the milk (R3) growth stage, and all treatments contained a nonionic surfactant (Preference) at a rate of 0.25% v/v.

[y] Tar spot stromata was visually assessed as a percentage (0–100%) of leaf area on five plants in each plot at the ear leaf on September 29.

[x] Tar spot chlorosis and necrosis symptoms were visually assessed as a percentage (0–100%) of leaf area on five plants in each plot at the ear leaf on September 29.

[w] Canopy green was visually assess as the percentage (0–100%) in the plot on October 7.

[v] Yields were adjusted to 14.5 % moisture and to account for wind damage that caused stand loss in plots where it occurred at harvest on November 4.

[u] Means followed by the same letter are not significantly different based on Fisher's Least Significant Difference test (LSD; α=0.05).

EVALUATION OF XYWAY 3D SYSTEM FOR TAR SPOT OF CORN IN NORTHWESTERN INDIANA, 2020 (COR20-30.PPAC)

S. Shim, J. D. Ravellette, and D. E. P. Telenko, Department of Botany and Plant Pathology, Purdue University West Lafayette, Indiana 47907-2054

CORN (*ZEA MAYS* W2585SSRIB)

Tar spot, *Phyllachora maydis*

A trial was established at the Pinney Purdue Agricultural Center (PPAC) in Porter County, Indiana. The experiment was a randomized complete block design with four replications. Plots were 10 feet wide and 30 feet long and consisted of four rows, and the two center rows were used for evaluation. The previous crop was corn. Corn hybrid W2585SSRIB was planted in 30-inch row spacing at a rate of 34,000 seeds/acre on June 7. Standard practices for nonirrigated corn production in Indiana were followed. In-furrow applications were applied at planting using a Kincaid planter. All foliar fungicide applications were applied at 15 gal/acre and 40 psi using a Lee self-propelled sprayer equipped with a 10-foot boom, fitted with six TJ-VS 8002 nozzles spaced 20 inches apart, at 3.6 mph. Fungicides were applied on June 7 at planting, July 2 at V6, and August 7 at the tassel/silk (VT/R1) growth stages. Disease ratings were assessed on September 29 at maturity (R6) growth stage. Tar spot was rated by visually assessing the percentage of stromata and the percentage of symptomatic tissues (chlorosis and necrosis) per leaf on five plants in each plot at the ear leaf. Values for each plot were averaged before analysis. The two center rows of each plot were harvested on November 3, and yields were adjusted to 15.5% moisture. Data were subjected to mixed model analysis of variance (SAS 9.4, 2019), and means were compared using Fisher's Least Significant Difference test (LSD; α=0.05).

In 2020, weather conditions were favorable for disease. Tar spot was the most prominent disease in the trial and reached moderate severity. The Xyway 3D followed by (fb) Lucento, Topguard fb Lucento, Lucento, and Veltyma programs significantly reduced the severity of tar spot stromata over the nontreated control on September 29 (Table 33). No treatments reduced tar spot chlorotic and necrotic symptoms over the nontreated control. Xyway 5.9 fl oz in-furrow fb Lucento and Veltyma were the greenest plots on October 7 but were not significantly different from Topguard fb Lucento. There was no significant difference between treatments for harvest moisture and test weight. Xyway fb Lucento, Topguard fb Lucento, Lucento, and Veltyma increased yield over the nontreated control but were not significantly different from the other fungicide programs.

TABLE 33. *Effect of Fungicide Treatment on Tar Spot, Canopy Greenness, and Corn Yield*

TREATMENT, RATE/ACRE, AND TIMING[z]	TAR SPOT % SEVERITY[y]	TAR SPOT % CHLOR/NEC[x]	CANOPY GREEN[w] %	HARVEST MOISTURE %	TEST WEIGHT LB/BU	YIELD[v] BU/ACRE
Nontreated control	13.7 ab	14.9 bcd	27.5 c	24.1	53.1	179.5 c
Xyway 3D 5.9 fl oz in-furrow	18.0 a	24.3 ab	27.5 c	21.6	53.7	190.5 abc
Xyway 3D 11.8 fl oz in-furrow	13.8 ab	18.0 abc	31.3 bc	23.3	52.2	187.9 abc
Xyway 3D 5.9 fl oz in-furrow fb Lucento 4.17 SC 5.0 fl oz at VT[27]	6.2 c	9.9 cd	38.8 a	23.2	52.1	195.7 a
Headline EC 6.9 fl oz in-furrow	15.6 ab	18.3 abc	31.3 bc	22.7	52.4	187.7 abc
Topguard EQ 5.0 fl oz at V6	6.3 c	12.7 cd	36.3 ab	24.7	52.3	191.6 ab
Delaro 325 SC 4.0 fl oz at V6	12.4 b	26.5 a	27.5 c	22.8	52.4	182.4 bc
Lucento 4.17 SC 5.0 fl oz at VT	5.6 c	5.0 d	36.3 ab	23.9	52.3	196.8 a
Veltyma 3.34 S 7.0 fl oz at VT	2.0 c	8.2 cd	41.3 a	23.6	51.9	199.3 a
P-value[u]	<.0001	0.0084	0.0003	0.1427	0.0550	0.0346

[z] Fungicide treatments were applied on September 2 at the milk (R3) growth stage, and all treatments contained a nonionic surfactant (Preference) at a rate of 0.25% v/v. fb = followed by.

[y] Tar spot stromata was visually assessed as a percentage (0–100%) of leaf area on five plants in each plot at the ear leaf on September 29.

[x] Tar spot chlorosis and necrosis symptoms were visually assessed as a percentage (0–100%) of leaf area on five plants in each plot at the ear leaf on September 29.

[w] Canopy green was visually assessed as the percentage (0–100%) in the plot on October 7.

[v] Yields were adjusted to 15.5% moisture at harvest on November 3.

[u] Means followed by the same letter are not significantly different based on Fisher's Least Significant Difference test (LSD; α=0.05).

FUNGICIDE EVALUATION FOR TAR SPOT OF CORN IN NORTHWESTERN INDIANA, 2020 (COR20-31.PPAC)

S. Shim, J. D. Ravellette, and D. E. P. Telenko, Department of Botany and Plant Pathology, Purdue University West Lafayette, Indiana 47907-2054

CORN (*ZEA MAYS* W2585SSRIB)

Tar spot, *Phyllachora maydis*

A trial was established at the Pinney Purdue Agricultural Center (PPAC) in Porter County, Indiana. The experiment was a randomized complete block design with four replications. Plots were 10 feet wide and 30 feet long and consisted of four rows, and the two center rows were used for evaluation. The previous crop was corn. Corn hybrid W2585SSRIB was planted in 30-inch row spacing at a rate of 34,000 seeds/acre on June 9. Standard practices for nonirrigated grain corn production in Indiana were followed. All fungicide applications were applied at 15 gal/acre and 40 psi using a Lee self-propelled sprayer equipped with a 10-foot boom, fitted with six TJ-VS 8002 nozzles spaced 20 inches apart, at 3.6 mph. Fungicides were applied on August 8 at tassel/silk (VT/R1) growth stage. Disease ratings were assessed on October 7 at maturity (R6) growth stage. Tar spot was rated by visually assessing the percentage of stromata and the percentage of symptomatic tissues (chlorosis and necrosis) per leaf on five plants in each plot at the ear leaf. Values for each plot were averaged before analysis. The two center rows of each plot were harvested on November 6, and yields were adjusted to 15.5% moisture. Data were subjected to mixed model analysis of variance (SAS 9.4, 2019), and means were compared using Fisher's Least Significant Difference test (LSD; α=0.05).

In 2020, weather conditions were favorable for disease. Tar spot was the most prominent disease in the trial and reached moderate severity. All fungicide applications significantly reduced the severity of tar spot stromata and chloric and necrotic symptoms on the ear leaf on October 7 over the nontreated control (Table 34). All fungicide programs increased canopy greenness on October 7. No significant differences were found between treatments and the nontreated control for harvest moisture, test weight, and corn yield.

TABLE 34. *Effect of Fungicide Treatment on Tar Spot, Canopy Greenness, and Corn Yield*

TREATMENT AND RATE/ACRE[z]	TAR SPOT % STROMA[y]	TAR SPOT % CHLO/ NECR[x]	CANOPY GREEN[w] %	HARVEST MOISTURE %	TEST WEIGHT LB/BU	YIELD[v] BU/ACRE
Nontreated control	23.0 a	30.8 a	41.3 e	25.8	49.6	205.7
Topguard EQ 4.29 SC 5.0 fl oz	4.2 b	2.9 b	62.5 bcd	26.3	49.0	197.8
Lucento 4.17 SC 5.0 fl oz	5.0 b	3.9 b	55.0 d	25.9	49.2	214.5
Lucento 4.17 SC 5.0 fl oz + Quadris 6.0 fl oz	4.0 bc	3.8 b	60.0 bcd	26.0	49.3	208.9
Veltyma 3.34 S 7.0 fl oz	1.8 c	1.6 b	75.0 a	25.5	50.4	204.3
Miravis Neo 2.5 SE 13.7 fl oz	3.9 bc	3.5 b	56.3 cd	26.0	49.0	212.5
Trivapro 2.21 SE 13.7 fl oz	4.5 b	1.5 b	63.8 bc	25.9	49.4	208.2
Delaro 325 SC 8.0 fl oz	4.0 bc	1.8 b	66.3 b	25.5	49.3	205.8
P-value[u]	<.0001	<.0001	<.0001	0.2058	0.0130	0.8053

[z] Fungicides were applied August 8 at the tassel/silk (VT/R1) growth stage, and all treatments contained a nonionic surfactant (Preference) at a rate of 0.25% v/v.

[y] Tar spot stromata was visually assessed as a percentage (0–100%) of leaf area on five plants in each plot at the ear leaf on October 7.

[x] Tar spot chlorosis and necrosis symptoms were visually assessed as a percentage (0–100%) of leaf area on five plants in each plot at the ear leaf on October 7.

[w] Canopy green was visually assessed as a percentage (0–100%) on October 7.

[v] Yields were adjusted to 15.5% moisture at harvest on November 6.

[u] Means followed by the same letter are not significantly different based on Fisher's Least Significant Difference test (LSD; α=0.05).

FUNGICIDE COMPARISON FOR WHITE MOLD IN SOYBEAN IN NORTHWESTERN INDIANA, 2020 (SOY20-02.PPAC)

D. E. P. Telenko, J. D. Ravellette, and S. Shim, Department of Botany and Plant Pathology, Purdue University West Lafayette, Indiana 47907-2054

SOYBEAN (*GLYCINE MAX* P25A27X)

White mold, *Sclerotinia sclerotiorum*

A trial was established at the Pinney Purdue Agricultural Center (PPAC) in Porter County, Indiana. The experiment was a randomized complete block design with four replications. Plots were 6.7 feet wide and 30 feet long and consisted of four rows, and the two center rows were used for evaluation. The previous crop was corn. Standard practices for soybean production in Indiana were followed. Soybean cultivar P34A79X was planted in 20-inch row spacing at a rate of 8 seeds/foot on June 5. Inoculum of *S. sclerotiorum* was applied on the seedbed at 1.25 g/foot at planting. The field was overhead irrigated weekly at 1 inch unless weekly rainfall was 1 inch or higher to encourage disease. All fungicide applications were applied at 15 gal/acre and 40 psi using a Lee self-propelled sprayer equipped with a 10-foot boom, fitted with six TJ-VS 8002 nozzles spaced 20 inches apart, at 3.6 mph. Fungicides were applied on July 20 at the beginning bloom (R1) growth stage and August 5 at the beginning pod (R3) growth stage. Disease ratings were assessed on August 31 and September 10 at the beginning seed (R5) and full seed (R6) growth stages, respectively. White mold disease was assessed by counting the number of plants in each plot with symptoms. The two center rows of each plot were harvested on November 2, and yields were adjusted to 13% moisture. Data were subjected to mixed model analysis of variance (SAS 9.4, 2019), and means were compared using Fisher's Least Significant Difference test (LSD; α=0.05).

In 2020, very little disease developed in plots. There were no significant differences between fungicide treatments and the nontreated control for all disease ratings on August 31 and September 10 (Table 35). Aproach plus Aproach Prima program application at R1 followed by (fb) R1 +14d, Miravis Neo at R1, A21573 at R1, A21573 at R1 fb R1+14d, Aproach Prima at R1 fb R1+14d, Aproach at R1 fb R1+14d, and Revytek at R1 resulted in the greenest canopies and lowest defoliation on September 24. There was no significant effect of treatment on moisture, test weight, or soybean yield.

TABLE 35. *Effect of Fungicide on White Mold Incidence, Canopy Greenness, and Yield of Soybean*

TREATMENT, RATE/ACRE, AND TIMING[z]	WHITE MOLD[y]	CANOPY GREEN %[w]	DEFOLIATION %[x]	HARVEST MOISTURE %	TEST WEIGHT LB/BU	YIELD[v] BU/ACRE
Nontreated control	2.8	46.3 de	8.8 a	12.5	56.7	67.5
Contans 4.0 lb at plant	3.0	46.3 de	8.8 a	12.0	55.8	67.5
Double Nickel 2.0 qt at R1	3.8	46.3 de	7.5 ab	12.3	55.9	67.4
Contans 4.0 lb at planting fb Double Nickel 2.0 qt at R1	1.3	46.3 de	8.8 a	12.1	56.0	69.3
Miravis Neo 20.8 fl oz at R1	4.0	56.3 a	5.0 b	12.1	55.8	69.1
A21573C 13.7 fl oz at R1	2.3	53.8 abc	5.0 b	12.0	56.0	69.7
A21573C 13.7 fl oz at R1 fb A21573C 13.7 fl oz at R3	2.3	55.0 ab	5.0 b	12.0	55.9	70.9
Aproach Prima 6.8 fl oz at R1 fb Aproach Prima 6.8 fl oz at R1 + 14 d	3.0	50.0 a-e	5.0 b	12.0	55.9	70.8
Aproach 3.0 fl oz + Aproach Prima 6.8 fl oz at R1 fb Aproach 3.0 fl oz + Aproach Prima 6.8 fl oz at R1 + 14 d	0.8	56.3 a	5.0 b	12.0	56.0	72.0
Aproach 9.0 fl oz at R1 fb Aproach 9.0 fl oz at R1 + 14 d	1.8	52.5 a-d	5.0 b	12.0	56.1	70.1
Endura 12.5 oz at R1	1.5	45.0 e	7.5 ab	12.0	56.0	69.2
Priaxor Xemium 8.0 fl oz at R1	4.5	48.8 b-e	7.5 ab	12.0	55.7	69.3
Acropolis 23.0 fl oz at R1	5.0	47.5 cde	8.8 a	12.0	56.3	67.1
Revytek 15.0 fl oz at R1	5.8	51.3 a-e	5.0 b	12.1	56.0	69.2
P-value[u]	0.1189	0.0032	0.0198	0.5582	0.3739	0.2175

[z] Fungicide treatments were applied on July 20 at the R1 (beginning bloom) growth stage and August 5 at the R3 (beginning pod) growth stage, respectively. All treatments contained a nonionic surfactant (Induce) at a rate of 0.12% v/v. All plots were inoculated with *S. sclerotiorum*. fb = followed by.

[y] White mold disease was assessed by counting the number of plants per plot with symptoms on September 10.

[x] Green was visually assessed the percentage (0–100%) in the plot on September 24.

[w] Defoliation = percentage of leaf loss in plot.

[v] Yields were adjusted to 13% moisture at harvest on November 2.

[u] Means followed by the same letter are not significantly different based on Fisher's Least Significant Difference test (LSD; α=0.05).

COMPARISON OF SEED TREATMENTS FOR SUDDEN DEATH SYNDROME IN SOYBEAN IN INDIANA, 2020 (SOY20-03.PPAC)

S. Shim, J. D. Ravellette, and D. E. P. Telenko, Department of Botany and Plant Pathology, Purdue University West Lafayette, Indiana 47907-2054

SOYBEAN (*GLYCINE MAX* KSC33RX70C)

Sudden death syndrome, *Fusarium virguliforme*
Soybean cyst nematode, *Heterodera glycines*

A trial was established at the Pinney Purdue Agricultural Center (PPAC) in Porter County, Indiana. The experiment was a randomized complete block design with four replications. Plots were 10 feet wide and 30 feet long and consisted of four rows, and the two center rows were used for evaluation. The previous crop was corn. Standard practices for soybean production in Indiana were followed. Soybean cultivar KSC33RX70C was planted in 30-inch row spacing at a rate of 8 seeds/foot on June 6. All plots were inoculated with isolates of *Fusarium virguliforme* within the seedbed at 1.25 g/foot on June 6. Seed treatments were applied on seeds before planting. Disease ratings were assessed on August 31 and September 9 at the beginning seed (R5) and full seed (R6) growth stages, respectively. Sudden death syndrome (SDS) in each plot was rated for disease incidence (DI) and disease severity (DS). Disease incidence refers to the percentage of plants with disease symptoms, and disease severity (DS) was rated using a 1–9 scale where 1 refers to low disease pressure and 9 refers to premature death of the plant. SDS index was then calculated using the equation $DX = (DI \times DS)/9$. The two center rows of each plot were harvested on November 2, and yields were adjusted to 13% moisture. Data were subjected to mixed model analysis of variance (SAS 9.4, 2019), and means were compared using Fisher's Least Significant Difference test (LSD; α=0.05).

In 2020, very little disease developed in plots. SDS was the most prominent disease in the trial but only reached low severity. Soybean cyst nematode egg count in spring soil samples from field site was 0–700 eggs/100 cc soil, a low to moderate range. There was no significant difference between seed treatments for all disease ratings on September 10 (Table 36). There were no significant differences between seed treatments for canopy green, defoliation, harvest moisture, test weight, and yield.

TABLE 36. *Effect of Seed Treatment on Sudden Death Syndrome and Soybean Yield*

TREATMENT[z]	SDS %DI[y]	SDS DS[y]	SDS INDEX[y]	CANOPY GREEN[x]	DEFOLIATION[w] %	HARVEST MOISTURE %	TEST WEIGHT LB/BU	YIELD BU/ACRE
Base	3.8	0.8	0.4	38.8	7.5	12.6	56.4	65.2
ILEVO	2.8	1.0	0.3	43.8	6.3	12.5	56.5	64.6
BAS780 06 F	5.3	1.3	0.9	43.8	6.3	12.6	56.6	64.3
Saltro	5.0	1.0	0.6	45.0	6.3	12.6	56.6	63.3
BIOst + Mertect + Heads Up	5.3	1.3	0.9	42.5	8.8	12.5	56.3	62.8
BASF494	5.0	1.3	0.7	43.8	5.0	12.5	56.6	62.5
ILEVO + 725AWS	5.0	1.0	0.6	42.5	5.0	12.6	56.3	65.9
P-value[u]	*0.7197*	*0.7268*	*0.7309*	*0.2253*	*0.5161*	*0.6913*	*0.3764*	*0.6391*

[z] Seed treatments were preapplied to the seeds before planting. All plots were inoculated with isolates of *Fusarium virguliforme* within the seedbed at 1.25 g/foot on June 6.

[y] Sudden death syndrome (SDS) in each plot was rated for disease incidence (DI) and disease severity (DS) on September 10. DI refers to the percentage of plants with disease symptoms, and DS was rated using a 1–9 scale where 1 refers to low disease pressure and 9 refers to premature death of the plant. SDS index was then calculated using the equation DX = (DI x DS)/9.

[x] Canopy green was visually assessed as the percentage (0–100%) in the plot on September 24.

[w] Defoliation = percentage of leaf loss in plot on September 24.

[v] Yields were adjusted to 13% moisture at harvest on November 2.

[u] Means followed by the same letter are not significantly different based on Fisher's Least Significant Difference test (LSD; α=0.05).

EVALUATION OF SEED TREATMENTS FOR SOYBEAN SUDDEN DEATH SYNDROME IN NORTHWESTERN INDIANA, 2020 (SOY20-22.PPAC)

S. Shim, J. D. Ravellette, and D. E. P. Telenko, Department of Botany and Plant Pathology, Purdue University West Lafayette, Indiana 47907-2054

SOYBEAN (*GLYCINE MAX* GH3319E3)

Sudden death syndrome, *Fusarium virguliforme*
Soybean cyst nematode, *Heterodera glycines*

A trial was established at the Pinney Purdue Agricultural Center (PPAC) in Porter County, Indiana. The experiment was a randomized complete block design with four replications. Plots were 10 feet wide and 30 feet long and consisted of four rows, and the two center rows were used for evaluation. The previous crop was corn. Standard practices for soybean production in Indiana were followed. Soybean cultivar GH3319E3 was planted in 30-inch row spacing at a rate of 8 seeds/foot on June 6. All plots were inoculated with *Fusarium virguliforme* at 1.25 g/foot within the seedbed at planting. Seed treatments were preapplied to the seeds before planting. Disease ratings were assessed on September 10 at full seed (R6) growth stage. Sudden death syndrome (SDS) in each plot was rated for disease incidence (DI) and disease severity (DS). Disease incidence refers to the percentage of plants with disease symptoms, and disease severity (DS) was rated using a 1–9 scale where 1 refers to low disease pressure and 9 refers to premature death of the plant. SDS index was then calculated using the equation DX = (DI x DS)/9. The two center rows of each plot were harvested on November 2, and yields were adjusted to 13% moisture. Data were subjected to mixed model analysis of variance (SAS 9.4, 2019), and means were compared using Fisher's Least Significant Difference test (LSD; α=0.05).

In 2020, very little disease developed in plots. SDS was the most prominent disease but only reached low severity. Soybean cyst nematode egg count in spring soil samples ranged from 0 to 700 eggs/100 cc soil, a low to moderate range. No treatment differences were detected for SDS DI and SDS DS on September 10 (Table 37). ILeVO and BAS780 06 F seed treatments reduced SDS index over the nontreated control on September 10. There were no significant differences between seed treatments for harvest moisture, test weight, and yield of soybean.

TABLE 37. *Effect of Seed Treatment on Sudden Death Syndrome and Yield of Soybean*

TREATMENT[z]	SDS DI[y]	SDS DS[y]	SDS INDEX[y]	HARVEST MOISTURE %	TEST WEIGHT LB/BU	YIELD[x] BU/ ACRE
Base	6.3	2.0	1.3 a	12.8	55.9	62.8
ILEVO	3.8	0.8	0.5 b	12.8	55.8	61.6
BAS780 06 F	3.8	0.8	0.5 b	12.8	55.7	61.5
P-value[w]	0.2963	0.0805	0.0261	0.5787	0.8976	0.7791

[z] Seed treatments were applied to the seeds prior to planting.

[y] Sudden death syndrome (SDS) in each plot was rated for disease incidence (DI) and disease severity (DS) on September 10. Disease incidence refers to the percentage of plants with disease symptoms, and disease severity (DS) was rated using a 1–9 scale where 1 refers to low disease pressure and 9 refers to premature death of the plant. SDS index was then calculated using the equation DX = (DI x DS)/9.

[x] Yields were adjusted to 13% moisture at harvest on November 2.

[w] Means followed by the same letter are not significantly different based on Fisher's Least Significant Difference test (LSD; α=0.05).

COMPARISON OF FUNGICIDES FOR WHITE MOLD IN SOYBEAN IN NORTHWESTERN INDIANA, 2020 (SOY20-26.PPAC)

D. E. P. Telenko, J. D. Ravellette, and S. Shim, Department of Botany and Plant Pathology, Purdue University West Lafayette, Indiana 47907-2054

SOYBEAN (*GLYCINE MAX* P25A27X)

White mold, *Sclerotinia sclerotiorum*

A trial was established at the Pinney Purdue Agricultural Center (PPAC) in Porter County, Indiana. The experiment was a randomized complete block design with four replications. Plots were 6.7 feet wide and 30 feet long and consisted of four rows, and the two center rows were used for evaluation. The previous crop was corn. Standard practices for soybean production in Indiana were followed. Soybean cultivar P34A79X was planted in 20-inch row spacing at a rate of 8 seeds/foot on June 5. Inoculum of *S. sclerotiorum* was applied on the seedbed at 1.25 g/foot at planting. The field was overhead irrigated weekly at 1 inch unless weekly rainfall was 1 inch or higher to encourage disease. All fungicide applications were applied at 15 gal/acre and 40 psi using a Lee self-propelled sprayer equipped with a 10-foot boom, fitted with six TJ-VS 8002 nozzles spaced 20 inches apart, at 3.6 mph. Fungicides were applied on July 20 at the beginning bloom (R1) growth stage and August 5 at the beginning pod (R3) growth stage. Disease ratings were assessed on August 31 and September 10 at the beginning seed (R5) and full seed (R6) growth stages, respectively. White mold disease was assessed by counting the number of plants in each plot with symptoms. The two center rows of each plot were harvested on November 2, and yields were adjusted to 13% moisture. Data were subjected to mixed model analysis of variance (SAS 9.4, 2019), and means were compared using Fisher's Least Significant Difference test (LSD; $\alpha=0.05$).

In 2020, very little disease developed in plots. White mold was present in the trial but only reached low severity. There was no significant differences between fungicide treatments and the nontreated control for all disease ratings on August 31 and September 10 (Table 38). There was no significant effect of treatment on moisture, test weight, or soybean yield.

TABLE 38. *Effect of Fungicide on White Mold and Yield of Soybean*

TREATMENT, RATE/ACRE, AND TIMING[z]	WHITE MOLD[y] 31-AUGUST	WHITE MOLD[y] 10-SEPTEMBER	HARVEST MOISTURE %	TEST WEIGHT LB/BU	YIELD[x] BU/ACRE
Nontreated control	0.8	1.3	12.3	45.6	70.6
Propulse 400 SC 7.0 fl oz at R1	0.3	1.5	12.0	45.6	68.1
USF0411 458 SC 8.0 fl oz at R1	0.5	1.8	12.0	45.3	68.0
USF0411 458 SC 8.0 fl oz at R1 fb R3	0.3	2.3	11.9	45.2	68.0
P-value[w]	0.3272	0.8060	0.2881	0.5517	0.2190

[z] Fungicide treatments were applied on July 20 at the beginning bloom (R1) growth stage and August 5 at the beginning pod (R3) growth stage, respectively. All treatments contained a nonionic surfactant (Induce) at a rate of 0.12% v/v. All plots were inoculated with *S. sclerotiorum*. fb = followed by.

[y] White mold disease was assessed by counting the number of plants per plot with symptoms.

[x] Yields were adjusted to 13% moisture at harvest on November 2.

[w] Means followed by the same letter are not significantly different based on Fisher's Least Significant Difference test (LSD; α=0.05).

EVALUATION OF CULTIVARS AND SEED TREATMENT IN SOYBEAN IN NORTHWESTERN INDIANA, 2020 (SOY20-30.PPAC)

S. Shim, J. D. Ravellette, and D. E. P. Telenko, Department of Botany and Plant Pathology, Purdue University West Lafayette, Indiana 47907-2054

SOYBEAN (*GLYCINE MAX* P25A27X AND P24T76E)

Sudden death syndrome, *Fusarium virguliforme*
Soybean cyst nematode, *Heterodera glycines*

A trial was established at the Pinney Purdue Agricultural Center (PPAC) in Porter County, Indiana. The experiment was a randomized complete block design with four replications. Plots were 10 feet wide and 30 feet long and consisted of four rows, and the two center rows were used for evaluation. The previous crop was corn. Standard practices for soybean production in Indiana were followed. Soybean cultivar P25A27X (resistant) and P24T76E (susceptible) were planted in 30-inch row spacing at a rate of 8 seeds/foot on June 5. Seed treatments were applied on seeds before planting: resistant nontreated control, resistant ILEVO (0.15 mg/seed), resistant Saltro (standard rate), susceptible nontreated control, susceptible ILEVO (0.15 mg/seed), and susceptible Saltro (standard rate). Disease ratings were assessed on August 31 at the R5 (beginning pod/full pod) growth stage. Sudden death syndrome (SDS) in each plot was rated for disease incidence (DI) and disease severity (DS). Disease incidence refers to the percentage (0–100%) of plants with disease symptoms, and disease severity was rated using a 1–9 scale where 1 refers to low disease pressure and 9 refers to premature death of the plant. SDS index was then calculated using the equation $DX = (DI \times DS)/9$. The two center rows of each plot were harvested on November 2, and yields were adjusted to 13% moisture. Data were subjected to mixed model analysis of variance (SAS 9.4, 2019), and means were compared using Fisher's Least Significant Difference test (LSD; $\alpha=0.05$).

In 2020, very little disease developed in plots. SDS was the most prominent disease in the trial but only reached low severity. Soybean cyst nematode egg count in spring soil samples ranged from 0 to 500 eggs/100 cc soil, a low to moderate range. P25A27X (resistant) had significantly lower levels of SDS incidence and index over the susceptible cultivar, P24T76E (Table 39). There was no significant difference among seed treatments and nontreated control in each cultivar. There were no significant differences between seed treatments and cultivar for harvest moisture, test weight, and yield.

TABLE 39. *Effect of Fungicide on Sudden Death Syndrome and Soybean Yield*

TREATMENT AND VARIETY[z]	SDS DI[y]	SDS DS[y]	SDS INDEX[y]	HARVEST MOISTURE %	TEST WEIGHT LB/ BU	YIELD[x] BU/ ACRE
Nontreated control, P25A27X	1.8 b	0.5	0.2 b	11.9	55.3	57.9
ILEVO, P25A27X	2.0 b	0.5	0.2 b	11.9	55.3	64.1
Saltro, P25A27X	1.3 b	0.8	0.2 b	12.1	55.4	61.3
Nontreated control, P24T76E	10.0 a	1.3	1.3 a	11.8	55.2	58.1
ILEVO, P24T76E	10.0 a	1.5	1.4 a	11.7	55.3	56.5
Saltro, P24T76E	7.5 c	1.3	1.0 a	11.8	55.3	60.3
P-value[w]	0.0049	0.1701	<.0001	0.1695	0.9346	0.1348

[z] Seed treatments were applied to the seeds before planting on June 5: resistant nontreated control, resistant ILEVO (0.15 mg/ seed), resistant Saltro (standard rate), susceptible nontreated control, susceptible ILEVO (0.15 mg/seed), and susceptible Saltro (standard rate).

[y] Sudden death syndrome (SDS) in each plot was rated for disease incidence (DI) and disease severity (DS) on August 31. Disease incidence (DI) refers to the percentage (0–100%) of plants with disease symptoms, and disease severity (DS) was rated using a 1–9 scale where 1 refers to low disease pressure and 9 refers to premature death of the plant. SDS index was calculated using the equation DX = (DI x DS)/9.

[x] Yields were adjusted to 13% moisture at harvest on November 2.

[w] Means followed by the same letter are not significantly different based on Fisher's Least Significant Difference test (LSD;α=0.05).

SOUTHWEST PURDUE AGRICULTURAL CENTER (SWPAC)

EVALUATION OF FUNGICIDES FOR FOLIAR DISEASES ON CORN IN SOUTHWESTERN INDIANA, 2020 (COR20-18.SWPAC)

E. A. Dunham, J. D. Ravellette, and D. E. P. Telenko, Department of Botany and Plant Pathology, Purdue University West Lafayette, Indiana 47907-2054

CORN (ZEA MAYS P9998AM)

Southern rust, *Puccinia polysora*
Gray leaf spot, *Cercospora zeae-maydis*

A trial was established at the Southwest Purdue Agricultural Center (SWPAC) in Knox County, Indiana. The experiment was a randomized complete block design with four replications. Plots were 10 feet wide and 30 feet long and consisted of four rows, and the two center rows were used for evaluation. The previous crop was corn. Standard practices for grain corn production in Indiana were followed. Corn hybrid P9998AM was planted in 30-inch row spacing at a rate of 27,000 seeds/acre on May 22. All fungicide applications were applied at 15 gal/acre and 40 psi using a Lee self-propelled sprayer equipped with a 10-foot boom, fitted with six TJ-VS 8002 nozzles spaced 20 inches apart, at 3.6 mph. Fungicides were applied on July 18 at the tassel/silk (VT/R1) growth stage and August 6 at the milk (R3) growth stage. Disease ratings was assessed on September 1 at the dent (R5) growth stage. Disease severity was rated by visually assessing the percentage of symptomatic leaf area of the ear leaf on five leaves in each plot. Values for each plot were averaged before analysis. The two center rows of each plot were harvested on October 2, and yields were adjusted to 15.5% moisture. Data were subjected to mixed model analysis of variance (SAS 9.4, 2019), and means were compared using Fisher's Least Significant Difference test (LSD; α=0.05).

In 2020, weather conditions were favorable for disease. Southern rust and gray leaf spot were the most prominent diseases in the trial and reached moderate severity. All fungicides at both VT/R1 and R3 application timings significantly reduced southern rust and gray leaf spot as compared to the nontreated control (Table 40). No significant differences were detected between fungicide treatments and nontreated control for yield.

TABLE 40. *Effect of Fungicide on Foliar Diseases Severity and Corn Yield*

TREATMENT, RATE/ACRE, AND TIMING[z]	SR % SEVERITY[y]	GLS % SEVERITY[y]	HARVEST MOISTURE %	TEST WEIGHT LB/ BU	YIELD[x] BU/ ACRE
Nontreated control	16.9 a	5.1 a	15.8 c	58.5 ab	223.6
Lucento 4.17 SC 5.0 fl oz at VT/R1	3.0 c-g	0.5 de	16.9 ab	57.9 a-g	216.2
Trivapro 2.21 SE 13.7 fl oz at VT/R1	1.3 f	1.1 b-e	17.1 ab	57.5 efg	213.3
Miravis Neo 2.5 SE 13.7 fl oz at VT/R1	6.1 bc	0.6 de	16.8 ab	58.5 a	221.4
Veltyma 3.34 S 7.0 fl oz at VT/R1	3.6 b-g	0.2 e	16.5 bc	57.7 c-g	224.1
Delaro 325 SC 8.0 fl oz at VT/R1	4.4 b-f	0.6 de	16.8 ab	58.3 abc	225.6
Quilt Xcel 2.2 SE 10.5 fl oz at VT/R1	5.0 b-e	0.4 de	17.0 ab	58.2 a-d	218.0
Headline AMP 1.68 SC 10.0 fl oz at VT/R1	3.5 b-g	1.1 b-e	16.7 ab	58.1 a-e	204.3
Revytek 3.33 LC 8.0 fl oz at VT/R1	3.2 b-g	0.2 e	17.0 ab	57.9 a-g	215.3
Lucento 4.17 SC 5.0 fl oz at R3	1.2 fg	1.0 b-e	16.9 ab	58.2 a-d	216.5
Trivapro 2.21 SE 13.7 fl oz at R3	0.4 g	1.3 cde	17.2 ab	57.3 g	224.0
Miravis Neo 2.5 SE 13.7 fl oz at R3	2.5 d-g	1.8 bc	16.7 b	57.4 fg	214.7
Veltyma 3.34 S 7.0 fl oz at R3	1.5 fg	0.8 cde	17.0 ab	57.6 d-g	221.9
Delaro 325 SC 8.0 fl oz at R3	5.5 bcd	1.1 b-e	16.6 b	57.9 b-g	221.9
Quilt Xcel 2.2 SE 10.5 fl oz at R3	6.3 b	2.0 b	16.6 b	58.0 a-f	215.7
Headline AMP 1.68 SC 10 fl oz at R3	1.8 efg	2.1 b	16.7 b	57.9 b-g	225.6
Revytek 3.33 LC 8.0 fl oz at R3	1.4 fg	0.7 cde	17.5 a	57.5 efg	213.3
P-value[w]	<.0001	<.0001	0.0364	0.0049	0.2111

[z] Fungicide treatments were applied on July 18 at the tassel/silk (VT/R1) growth stage and August 6 at the milk (R3) growth stage, and all treatments contained a nonionic surfactant at a rate of 0.25% v/v.

[y] Disease severity was visually assessed as a percentage (0–100%) of symptomatic leaf area on ear leaf; five plants were assessed per plot and averaged before analysis. SR = southern rust, GLS = gray leaf spot.

[x] Yields were adjusted to 15.5% moisture at harvest on October 6.

[w] Means followed by the same letter are not significantly different based on Fisher's Least Significant Difference test (LSD; α=0.05).

EVALUATION OF FUNGICIDES FOR FOLIAR DISEASES ON SOYBEAN IN SOUTHWESTERN INDIANA, 2020 (SOY20-18.SWPAC)

N. Piñeros-Guerrero, J. D. Ravellette, and D. E. P. Telenko, Department of Botany and Plant Pathology, Purdue University West Lafayette, Indiana 47907-2054

SOYBEAN (*GLYCINE MAX* P32A87L)

Septoria brown spot, *Septoria glycines*
Cercospora leaf blight, *Cercospora kikuchii/C. flagellaris*

A trial was established at the Southwest Purdue Agricultural Center (SWPAC) in Knox County, Indiana. The experiment was a randomized complete block design with four replications. Plots were 10 feet wide and 30 feet long and consisted of four rows, and the two center rows were used for evaluation. The previous crop was corn. Standard practices for soybean production in Indiana were followed. Soybean cultivar P32A87L was planted in 30-inch row spacing at a rate of 150,000 seeds/acre on May 26. All fungicide were applied at 15 gal/acre and 40 psi using a Lee self-propelled sprayer equipped with a 10-foot boom, fitted with six TJ-VS 8002 nozzles spaced 20 inches apart, at 3.6 mph. Fungicides were applied on July 24 at the beginning pod (R3) growth stage. Disease ratings were assessed on August 19 at the beginning seed (R5) growth stage. Cercospora leaf blight (CLB), and Septoria brown spot (SBS) were rated for disease severity by visually assessing the percentage of symptomatic leaf area in the upper and lower canopies, respectively. The two center rows were harvested on September 30, and yields were adjusted to 13% moisture. Data were subjected to mixed model analysis of variance (SAS 9.4, 2019), and means were compared using Fisher's Least Significant Difference test (LSD; α=0.05).

In 2020, weather conditions were unfavorable for soybean disease. SBS and CLB were the most prominent diseases in the trial and reached low severity. No significant treatment differences were detected for SBS and CLB severity, soybean harvest moisture, test weight, and yield (Table 41).

TABLE 41. *Effect of Fungicide on Foliar Disease Severity and Soybean Yield*

TREATMENT AND RATE/ACRE [z]	CLB % SEVERITY[y]	SBS % SEVERITY[y]	HARVEST MOISTURE %	TEST WEIGHT LB/BU	YIELD [x] BU/ACRE
Nontreated control	0.00	0.33	11.3	55.7	60.6
Preemptor 3.22 SC 5.0 fl oz	0.00	0.08	11.5	55.3	59.1
Topguard EQ 5.0 fl oz	0.00	0.05	11.8	55.7	60.4
Quadris Top SBX 7.0 fl oz	0.00	0.05	11.4	55.4	59.5
Lucento 4.17 SC 5.0 fl oz	0.00	0.03	11.7	55.6	60.2
Miravis Top 1.67 SC 13.7 fl oz	0.03	0.00	11.8	55.1	64.1
Priaxor 4.17 SC 4.0 fl oz	0.00	0.05	11.2	55.3	59.6
Trivapro 2.21 SE 13.7 fl oz	0.00	0.05	11.1	55.5	54.3
Delaro 325 SC 8.0 fl oz	0.00	0.03	11.7	55.4	60.0
Headline AMP 1.68 SC 10.0 fl oz	0.00	0.03	11.7	55.7	61.7
Veltyma 3.34 S 7.0 fl oz	0.00	0.00	11.7	55.7	57.4
Revytek 3.33 LC 8.0 fl oz	0.00	0.10	11.8	55.4	62.1
P-value[w]	*0.4671*	*0.1284*	*0.7266*	*0.9121*	*0.3456*

[z] Fungicide treatments were applied on July 24 at the R3 growth stage, and all treatments contained a nonionic surfactant (Preference) at a rate of 0.25%.

[y] Foliar disease incidence was rated on a scale of 0–100% of plants within a plot with disease symptoms. CLB = Cercospora leaf blight, SBS = Septoria brown spot.

[x] Yields were adjusted to 13% moisture at harvest on September 30.

[w] Means followed by the same letter are not significantly different based on Fisher's Least Significant Difference Test (LSD; α=0.05).

FUSARIUM HEAD BLIGHT (FHB) UNIFORM FUNGICIDE TRIAL IN SOUTHWESTERN INDIANA, 2020 (WHT20-03.SWPAC)

C. Rocca Da Silva, J. D. Ravellette, S. Shim, and D. E. P. Telenko, Department of Botany and Plant Pathology, Purdue University West Lafayette, Indiana 47907-2054

WHEAT (*TRITICUM AESTIVUM* P25R40)

Fusarium head blight, *Fusarium graminearum*
Stagnospora leaf and glume blotch, *Phaeosphaeria nodorum*

A trial was established at the Southwest Purdue Agricultural Center (SWPAC) in Knox County, Indiana. The experiment was a randomized complete block design with four replications. Plots were 7.5 feet wide and 20 feet long and consisted of 12 rows spaced 7.5 inches apart, and the center of each plot was used for evaluation. The previous crop was corn. Prior to planting, the field was disked on October 10, October 15, and October 16, 2019. Nitrogen (46-0-0) at 50 lb/acre plus potash (0-0-60) at 200 lb/acre was applied on October 14, 2019. AMS (21-0-0-24) at 100 lb/acre plus boron 14.3% at 7 lb/acre plus nitrogen (46-0-0) at 200 lb/acre was applied on February 21, 2020. On October 19, 2019, wheat cultivar P25R40 was drilled at 7.5-inch spacing. Harmony Extra at 0.8 oz/acre plus AMS at 2 lb/acre plus NIS at 0.25% v/v was applied on April 1, 2020, for weed management. All fungicide applications were applied at 15 gal/acre and 40 psi using a Lee self-propelled sprayer equipped with a 10-foot boom, fitted with six TJ-VS 8002 nozzles spaced 20 inches apart and directed forward and backward at a 45-degree angle, at 3.6 mph. Fungicides were applied on May 7 at Feekes 10.3, on May 22 at Feekes 10.5.1, and on May 26 at Feekes 10.5.4. All plots were inoculated with a mixture of isolates of *Fusarium graminearum* endemic to Indiana on May 22. The spore suspension (50,000 spores/ml) was applied at 215 ml/plot at Feekes 10.5.1. Disease ratings were assessed on June 8. Fusarium head blight (FHB) disease incidence (DI) was measured as the number of infected heads out of 60 plants in each plot and calculated as a percentage. FHB disease severity (DS) was rated by visually assessing the percentage of the infected head. FHB index was calculated as (FHB % DI multiplied by average FHB % DS)/100 per plot. Disease severity of leaf blotch was rated by visually assessing the percentage of symptomatic tissue on five flag leaves per plot for leaf blotch. Values for each plot were averaged before analysis. The eight center rows of each plot were harvested with a Kincaid plot combine on June 25, and yields were adjusted to 13.5% moisture. Data were subjected to mixed model analysis of variance (SAS 9.4, 2019), and means were compared using Fisher's Least Significant Difference test (LSD; α=0.05).

In 2020, weather conditions were moderately favorable for FHB and leaf blotch. FHB was the most prominent disease in the trial. All fungicides reduced FHB % DI, FHB % DS, FHB index, and leaf blotch percent severity (Table 42). Miravis Ace applied at 10.5.4 resulted in the highest level of FHB % DI, FHB % DS, and FHB index when compared to all other fungicide programs. The concentration of mycotoxin deoxynivalenol (DON) was significantly reduced by all fungicide programs over the nontreated control. Caramba at 10.5.4 had the highest level of DON of the fungicide programs but was not different from Miravis Ace applied at Feekes 10.3 or 10.5.4 (Table 43). The percentage of Fusarium damaged kernels (FDK) was significantly reduced by all fungicide programs over the nontreated control. Harvest moisture and test weight were lowest in the nontreated control but were not different from treatments of Prosaro, Caramba, or Sphaerex applied at Feekes 10.5.1. There were no significant differences between treatments for wheat yield.

TABLE 42. *Effect of Fungicide on Fusarium Head Blight (FHB) and Leaf Blotch*

TREATMENT, RATE/ACRE, AND TIMING[z]	FHB % DI[y]	FHB % DS[y]	FHB INDEX[x]	LEAF BLOTCH % SEVERITY[w]
Nontreated control	95.0 a	38.8 a	37.0 a	16.0 a
Prosaro 421 SC 6.5 fl oz at 10.5.1	65.6 bc	15.1 bc	9.6 bc	2.5 b
Caramba 90 EC 13.5 fl oz at 10.5.1	58.1 bc	13.7 bc	8.0 c	3.1 b
Sphaerex 7.3 fl oz at 10.5.1	54.2 c	12.0 c	6.5 c	2.9 b
Prosaro PRO 400 SC 10.3 fl oz at 10.5.1	51.3 c	14.8 bc	7.5 c	2.6 b
Miravis Ace 5.2 SC 13.7 fl oz at 10.3	58.8 c	15.3 bc	9.2 bc	3.0 b
Miravis Ace 5.2 SC 13.7 fl oz at 10.5.1	60.0 bc	15.3 bc	9.3 bc	2.3 b
Miravis Ace 5.2 SC 13.7 fl oz at 10.5.4	74.6 b	17.1 b	13.2 b	3.8 b
Miravis Ace 5.2 SC 13.7 fl oz at 10.5.1 fb Prosaro 421 SC 6.5 fl oz at 10.5.4	57.9 c	11.3 c	6.5 c	3.4 b
Miravis Ace 5.2 SC 13.7 fl oz at 10.5.1 fb Caramba 90 EC 13.5 fl oz at 10.5.4	58.3 c	10.9 c	6.5 c	3.3 b
Miravis Ace 5.2 SC 13.7 fl oz at 10.5.1 fb Folicur 3.6 F 4.0 fl oz at 10.5.4	58.8 c	13.1 c	7.7 c	1.4 b
P-value[v]	0.0002	<.0001	<.0001	<.0001

[z] All treatments contained a nonionic surfactant (Preference) at a rate of 0.125% v/v. Plots were inoculated with *Fusarium graminearum* spore suspension (50,000 spores/ml) after the treatment at Feekes 10.5.1. Spore suspension was applied at 215 ml/plot with handheld sprayer on May 22. fb = followed by.

[y] FHB % DI = Fusarium head blight (FHB) percent disease incidence was measured as the number of infected heads out of 60 plants in each plot and calculated as a percentage. FHB % DS= FHB percent disease severity was rated by visually assessing the severity of the infected heads.

[x] FHB index was calculated as (FHB % DI/average FHB % DS)/100 per plot.

[w] Disease severity of leaf blotch was rated by visually assessing the percentage of symptomatic tissue on five flag leaves per plot on June 8

[v] Means followed by the same letter are not significantly different based on Fisher's Least Significant Difference test (LSD; α=0.05).

TABLE 43. *Effect of Fungicide on Deoxynivalenol (DON), Fusarium Damaged Kernels (FDK), and Wheat Yield*

TREATMENT, RATE/ACRE, AND TIMING[z]	DON[y] PPM	FDK[x] %	HARVEST MOISTURE %	TEST WEIGHT LB/ BU	YIELD[w] BU/ ACRE[28]
Nontreated control	4.8 a	11.38 a	12.5 d	56.9 e	74.5
Prosaro 421 SC 6.5 fl oz at 10.5.1	1.1 d	9.38 b	12.9 d	57.5 cde	84.7
Caramba 90 EC 13.5 fl oz at 10.5.1	3.1 b	8.92 b	12.8 d	57.5 b-e	76.0
Sphaerex 7.3 fl oz at 10.5.1	1.3 cd	8.00 b	12.7 d	57.2 de	79.9
Prosaro PRO 400 SC 10.3 fl oz at 10.5.1	0.9 d	7.88 b	13.3 c	58.5 abc	81.1
Miravis Ace 5.2 SC 13.7 fl oz at 10.3	2.5 bc	8.00 b	13.3 c	58.3 abc	81.5
Miravis Ace 5.2 SC 13.7 fl oz at 10.5.1	1.3 cd	8.00 b	13.8 ab	58.0 a-d	78.6
Miravis Ace 5.2 SC 13.7 fl oz at 10.5.4	2.1 bcd	8.75 b	13.5 bc	58.2 abc	78.7
Miravis Ace 5.2 SC 13.7 fl oz at 10.5.1 fb Prosaro 421 SC 6.5 fl oz at 10.5.4	0.9 d	7.63 b	14.0 a	59.0 a	82.0
Miravis Ace 5.2 SC 13.7 fl oz at 10.5.1 fb Caramba 90 EC 13.5 fl oz at 10.5.4	1.3 cd	7.63 b	14.0 a	58.5 ab	79.7
Miravis Ace 5.2 SC 13.7 fl oz at 10.5.1 fb Folicur 3.6 F 4.0 fl oz at 10.5.4	1.4 cd	7.63 b	13.8 ab	59.0 a	79.5
P-value[v]	<.0001	0.0059	<.0001	0.0019	0.2454

[z] All treatments contained a nonionic surfactant (Preference) at a rate of 0.125% v/v. Plots were inoculated with *Fusarium graminearum* spore suspension (50,000 spores/ml) after treatment at Feekes 10.5.1. Spore suspension was applied at 215 ml/plot with a handheld sprayer on May 22. fb = followed by.

[y] Analysis of the mycotoxin deoxynivalenol (DON) was completed by the University of Minnesota DON Testing Lab.

[x] FDK visual estimative = percentage of Fusarium damaged kernels out of a subsample taken from each plot.

[w] Yields were adjusted to 13.5% moisture at harvest on June 25.

[v] Means followed by the same letter are not significantly different based on Fisher's Least Significant Difference test (LSD;α=0.05).

FUSARIUM HEAD BLIGHT (FHB) INTEGRATED MANAGEMENT TRIAL IN SOUTHWEST INDIANA, 2020 (WHT20-04.SWPAC)

C. Rocca Da Silva, J. D. Ravellette, S. Shim, and D. E. P. Telenko, Department of Botany and Plant Pathology, Purdue University West Lafayette, Indiana 47907-2054

WHEAT (*TRITICUM AESTIVUM* P25R40 AND P25R61)

Fusarium head blight, *Fusarium graminearum*
Stagnospora leaf and glume blotch, *Phaeosphaeria nodorum*

A trial was established at the Southwest Purdue Agricultural Center (SWPAC) in Knox County, Indiana. The experiment was a randomized complete block design with four replications. Plots were 7.5 feet wide and 20 feet long and consisted of 12 rows spaced 7.5 inches apart, and the center of each plot was used for evaluation. The previous crop was corn. Prior to planting, the field was disked on October 10, October 15, and October 16, 2018. Nitrogen (46-0-0) at 50 lb/acre plus potash (0-0-60) at 200 lb/acre was applied on October 14, 2019. AMS (21-0-0-24) at 100 lb/acre plus boron 14.3% at 7 lb/acre plus nitrogen (46-0-0) at 200 lb/acre was applied on February 21, 2020. On October 19, 2019, wheat cultivars P25R40 and P25R61 were drilled at 7.5-inch spacing. Harmony Extra at 0.8 oz/acre plus AMS at 2 lb/acre plus NIS at 0.25% v/v was applied on April 1, 2020, for weed management. All fungicide applications were applied at 15 gal/acre and 40 psi using a Lee self-propelled sprayer equipped with a 10-foot boom, fitted with six TJ-VS 8002 nozzles spaced 20 inches apart and directed forward and backward at a 45-degree angle, at 3.6 mph. Fungicides were applied on May 7 at Feekes 10.3, on May 22 at Feekes 10.5.1, and on May 26 at Feekes 10.5.4. All plots were inoculated with a mixture of isolates of *Fusarium graminearum* endemic to Indiana on May 22. The spore suspension (50,000 spores/ml) was applied at 215 ml/plot at Feekes 10.5.1. Disease ratings were assessed on June 8. Fusarium head blight (FHB) disease incidence (DI) was measured as the number of infected heads out of 60 plants in each plot and calculated as a percentage. FHB disease severity (DS) was rated by visually assessing the percentage of the infected head. FHB index was calculated as (FHB % DI multiplied by average FHB % DS)/100 per plot. Disease severity of leaf blotch was rated by visually assessing the percentage of symptomatic tissue on five flag leaves per plot for leaf blotch. Values for each plot were averaged before analysis. The eight center rows of each plot were harvested with a Kincaid plot combine on June 25, and yields were adjusted to 13.5% moisture. Data were subjected to mixed model analysis of variance (SAS 9.4, 2019), and means were compared using Tukey-HSD (α=0.05).

In 2020, weather conditions were moderately favorable for FHB and leaf blotch. FHB was the most prominent disease in the trial. Main effects of cultivar and fungicide treatment are presented. FHB % DI, FHB % DS, FHB index, deoxynivalenol (DON), and percent Fusarium damages kernels (FDK) were lowest in the moderately resistant cultivar P25R61 (Tables 44 and 45). FHB % DI, FHB % DS, FHB index, and percent leaf blotch were reduced by all fungicide treatments over the nontreated controls on June 8 (Table 44). The concentration of DON was significantly reduced by all the fungicides over the nontreated, noninoculated control but not the nontreated, inoculated control (Table 45). Applications of Prosaro and Miravis Ace followed by Folicur had the lowest percent of FDK but was not significantly different from all other fungicide treatments.

Wheat test weight was higher in cultivar P25R40, and yields were highest in P25R40. Harvest moisture was significantly higher in treatments that included Miravis Ace, and test weight was highest with Miravis Ace followed by Folicur. Miravis Ace applied at Feekes 10.3, and the program of Miravis Ace followed by Folicur had increased yield over the nontreated controls, but these were not significantly different from the other fungicide programs.

TABLE 44. *Effect of Cultivar and Fungicide on Fusarium Head Blight (FHB) and Foliar Diseases in Wheat*

	FHB % DI[Y]	FHB % DS[Y]	FHB INDEX[X]	LEAF BLOTCH[W] %
Cultivar				
P25R40	73.3 a[V]	27.3 a	21.7 a	4.2
P25R61	38.6 b	9.7 b	4.0 b	3.8
Fungicide program				
Nontreated control, inoculated control	72.7 a	30.0 a	25.3 a	9.0 a
Prosaro 421 SC 6.5 fl oz at 10.5.1	52.9 b	15.6 b	9.6 b	1.6 b
Miravis Ace 5.2 SC 13.7 fl oz at 10.5.1	48.5 b	14.3 b	8.2 b	1.1 b
Miravis Ace 5.2 SC 13.7 fl oz at 10.3	37.3 b	13.2 b	6.0 b	1.0 b
Miravis Ace 5.2 SC 13.7 fl oz at 10.5.1 fb Folicur 4.0 fl oz at 10.5.3	48.4 b	12.4 b	7.0 b	0.8 b
Nontreated, noninoculated control	75.8 a	25.6 a	21.1 a	10.4 a
Cultivar P-value[V]	<.0001	<.0001	<.0001	0.6329
Treatment P-value	<.0001	<.0001	<.0001	<.0001
*Cul*Trt P*-value	0.8682	<.0001	<.0001	0.5770

[Z] Fungicide treatments were applied at Feekes 10.3, 10.5.1, and 10.5.3. All treatments contained a nonionic surfactant (Preference) at a rate of 0.125% v/v. All plots were inoculated with *Fusarium graminearum* spore suspension (50,000 spores/ml) after the treatment at Feekes 10.5.1 except for the nontreated, noninoculated control. Spore suspension was applied at 300 ml/plot on May 22. fb = followed by.

[Y] FHB % DI = Fusarium head blight (FHB) percent disease incidence was measured as the number of infected heads out of 60 plants in each plot and calculated as a percentage. FHB % DS = FHB percent disease severity was rated by visually assessing the severity of the infected heads.

[X] FHB index was calculated as (FHB % DI/average FHB % DS)/100 per plot.

[W] Disease severity of Stagnospora leaf and glume blotch was rated by visually assessing the percentage of symptomatic leaf tissue on five flag leaves per plot for leaf blotch and five heads per plot for glume blotch.

[V] Means followed by the same letter are not significantly different based on Tukey-HSD (α=0.05).

TABLE 45. *Effect of Cultivar and Fungicide on Deoxynivalenol (DON), Fusarium Damaged Kernels (FDK), and Yield of Wheat*

	DON[Y] PPM	FDK[X] %	HARVEST MOISTURE %	TEST WEIGHT LB/BU	YIELD[W] BU/A
Cultivar					
P25R40	2.674 a	8.4 a	13.4	57.7 a	78.8 b
P25R61	0.847 b	6.9 b	13.4	56.0 b	92.3 a
Fungicide program					
Nontreated control, inoculated control	1.994 b	8.4 ab	12.9 c	56.4 b	81.0 b
Prosaro 421 SC 6.5 fl oz at 10.5.1	1.196 b	6.8 c	13.2 b	53.7 ab	86.8 ab
Miravis Ace 5.2 SC 13.7 fl oz at 10.5.1	1.195 b	7.4 bc	13.8 a	57.1 ab	86.5 ab
Miravis Ace 5.2 SC 13.7 fl oz at 10.3	1.445 b	7.1 bc	13.8 a	57.2 ab	89.3 a
Miravis Ace 5.2 SC 13.7 fl oz at 10.5.1 fb Folicur 4.0 fl oz at 10.5.3	1.271 b	6.8 c	14.1 a	57.7 a	88.2 a
Nontreated, noninoculated control	3.463 a	9.3 a	12.8 c	56.1 b	81.5 b
Cultivar P-value[v]	<.0001	<.0001	0.3919	<.0001	<.0001
Treatment P-value	0.0002	<.0001	<.0001	0.0042	0.0004
Cul*Trt P-value	0.0376	0.0001	0.6025	0.3097	0.6860

[z] Fungicide treatments were applied at Feekes 10.3, 10.5.1, and 10.5.3. All treatments contained a nonionic surfactant (Preference) at a rate of 0.125% v/v. All plots were inoculated with *Fusarium graminearum* spore suspension (50,000 spores/ml) after the treatment at Feekes 10.5.1 except for the nontreated, noninoculated control. Spore suspension was applied at 300 ml/plot on May 22. fb = followed by.

[y] Analysis of the mycotoxin deoxynivalenol (DON) was completed by the University of Minnesota DON Testing Lab.

[x] FDK = percentage of Fusarium damaged kernels.

[w] Yields were adjusted to 13.5% moisture at harvest on June 25.

[v] Means followed by the same letter are not significantly different based on Tukey-HSD (α=0.05).

DAVIS PURDUE AGRICULTURAL CENTER (DPAC)

FIELD-SCALE FUNGICIDE TIMING COMPARISON FOR FOLIAR DISEASES ON CORN IN CENTRAL INDIANA, 2020 (COR20-09.DPAC)

C. Rocca Da Silva, S. Shim, J. D. Ravellette, and D. E. P. Telenko, Department of Botany and Plant Pathology, Purdue University West Lafayette, Indiana 47907-2054

CORN (*ZEA MAYS* P0157AM)

Southern rust, *Puccinia polysora*
Northern corn leaf blight, *Setosphaeria turcica*
Gray leaf spot, *Cercospora zeae-maydis*

A trial was established at the Davis Purdue Agricultural Center (DPAC) in Randolph County, Indiana. The experiment was a randomized complete block design with four replications. Plots were 30 feet wide and 500 feet long and consisted of 12 rows, and the two center rows were used for evaluation. The previous crop was soybean. Standard practices for nonirrigated soybean production in Indiana were followed. Corn hybrid P0157AM was planted in 30-inch row spacing at a rate of 30,000 seeds/acre on May 6. All fungicide applications were applied at 20 gal/acre and 40 psi using an Apache 720 sprayer. All fungicides were applied on June 25 at V6, on July 10 at V10, and on July 30 at tassel/silk (VT/R1) growth stages. Southern rust (SR), northern corn leaf blight (NCLB), and gray leaf spot (GLS) were assessed on September 11 at the maturity (R6) growth stage. Disease severity was rated by visually assessing the percentage of symptomatic leaf area on 10 plants in each plot at the ear leaf. Values for each plot were averaged before analysis. Data were subjected to mixed model analysis of variance (SAS 9.4, 2019), and means were compared using Fisher's Least Significant Difference test (LSD; α=0.05).

In 2020, southern rust (SR), northern corn leaf blight (NCLB), and gray leaf spot (GLS) were the most prominent diseases in the trial. Application timings (V10 and VT/R1) of Delaro reduced GLS on the ear leaf (Table 46). There was no difference for SR and NCLB on the ear leaf. There was no significant between treatments for canopy greenness, moisture, and yield of corn.

TABLE 46. *Effect of Fungicide on Foliar Disease Severity, Canopy Greenness, and Corn Yield*

TREATMENT, RATE/ACRE, AND TIMING[z]	SR % SEVERITY[y]	NCLB % SEVERITY[y]	GLS % SEVERITY[y]	CANOPY GREEN[x]	HARVEST MOISTURE %	YIELD[w] BU/ACRE
Nontreated control	0.6	1.4	3.3 a	61.3	17.3	210.3
Delaro 325 SC 8.0 fl oz at V6	0.7	1.4	2.9 a	58.8	17.5	213.0
Delaro 325 SC 8.0 fl oz at V10	0.3	0.6	1.0 b	60.0	17.4	200.9
Delaro 325 SC 8.0 fl oz at VT/R1	0.3	1.4	1.0 b	66.3	17.6	205.5
P-value	*0.2089*	*0.425*	*0.0073*	*0.5062*	*0.0597*	*0.1361*

[z] Fungicide treatments were applied on June 25 at V6, on July 10 at V10, and on July 30 at tassel/silk (VT/R1) growth stages. All treatments contained a nonionic surfactant (Preference) at a rate of 0.25% v/v.

[y] Disease severity was visually assessed as a percentage (0–100%) of symptomatic leaf area on ear leaf on September 11. Ten leaves assessed per plot and averaged. SR = southern rust, NCLB = northern corn leaf blight, GLS = gray leaf spot.

[x] Canopy green was visually assessed as a percentage (0–100%) of crop canopy on September 11.

[w] Yields were adjusted to 15.5% moisture at harvest on October 28.

[v] Means followed by the same letter are not significantly different based on Fisher's Least Significant Difference test (LSD; α=0.05).

FIELD-SCALE FUNGICIDE TIMING COMPARISON FOR FOLIAR DISEASES ON SOYBEAN IN CENTRAL INDIANA, 2020 (SOY20-10.DPAC)

C. Rocca Da Silva, S. Shim, J. D. Ravellette, and D. E. P. Telenko, Department of Botany and Plant Pathology, Purdue University West Lafayette, Indiana 47907-2054

SOYBEAN (*GLYCINE MAX* P32A87L)

Septoria brown spot, *Septoria glycines*
Cercospora leaf blight, *Cercospora kikuchii/C. flagellaris*

A trial was established at the Davis Purdue Agricultural Center (DPAC) in Randolph County, Indiana. The experiment was a randomized complete block design with four replications. Plots were 30 feet wide and 460 feet long and consisted of 24 rows, and the two center rows were used for evaluation. The previous crop was corn. Standard practices for nonirrigated soybean production in Indiana were followed. Soybean cultivar P32A87LL was planted in 15 inches row spacing at a rate of 150,000 seeds/acre on May 12. All fungicide applications were applied at 20 gal/acre and 40 psi using an Apache 720 sprayer with Trimble CFX monitor for rate and section control and RTK guidance. Fungicides were applied on June 25 at the V4, July 30 at the beginning pod (R3), and August 21 at the beginning seed (R5) growth stages. Disease ratings were assessed on September 11 at the full seed (R6) growth stage. Septoria brown spot (SBS) and Cercospora leaf blight (CLB) were rated for disease severity by visually assessing the percentage of symptomatic leaf area in the upper and lower canopies, respectively. Data were subjected to mixed model analysis of variance (SAS 9.4, 2019), and means were separated using Fisher's Least Significant Difference test (LSD; α=0.05).

In 2020, very little disease developed in plots. SBS and CLB were the most prominent diseases but only reached low severity. Delaro applied at the R5 growth stage reduced SBS in the upper and lower canopies compared to the nontreated control on September 11 (Table 47). There was no significant difference for CLB in the upper canopy. There was no significant treatment effect on canopy green, defoliation, moisture, and corn yield.

TABLE 47. *Effect of Fungicide on Foliar Disease Severity, Canopy Greenness, Defoliation, and Soybean Yield*

TREATMENT, RATE/ACRE, AND TIMING[z]	SBS[y] % UPPER CANOPY	SBS[y] % LOWER CANOPY	CLB[y] % UPPER CANOPY	CANOPY GREEN[x] %	DEFOLIATION[w] %	MOISTURE %	YIELD[v] LB/A
Nontreated control	10.0 a	8.8 a	0.0	56.3	13.0	14.4	70.8
Delaro 325 SC 12.0 fl oz at V4	7.5 a	10.0 a	0.0	63.8	7.8	14.5	70.8
Delaro 325 SC 12.0 fl oz at R3	10.0 a	8.8 a	0.0	61.3	9.3	14.3	70.9
Delaro 325 SC 12.0 fl oz at R3	3.0 b	1.5 b	1.5	70.0	6.3	14.4	70.0
P-value[u]	0.0160	0.0050	0.2594	0.1787	0.4236	0.3903	0.9708

[z] Fungicide treatments were applied on June 25 at the V4, July 30 at the beginning pod (R3), and August 21 at the beginning seed (R5) growth stages. All treatments contained a nonionic surfactant (Preference) at a rate of 0.25% v/v.

[y] Foliar disease severity was visually rated on a scale of 0–100% of the upper and lower canopies with disease symptoms on September 11. SBS = Septoria brown spot, CLB = Cercospora leaf blight.

[x] Canopy green was visually assessed as a percentage (0–100%) of canopy on September 11.

[w] Defoliation = percentage of leaf loss in plot on September 11.

[v] Yields were adjusted to 13% moisture at harvest on October 5.

[u] Means followed by the same letter are not significantly different based on Fisher's Least Significant Difference test (LSD; α=0.05).

NORTHEAST PURDUE AGRICULTURAL CENTER (NEPAC)

FIELD-SCALE FUNGICIDE TIMING COMPARISON FOR FOLIAR DISEASES ON CORN IN NORTHEASTERN INDIANA, 2020 (COR20-10.NEPAC)

D. E. P. Telenko, J. D. Ravellette, and S. Shim, Department of Botany and Plant Pathology, Purdue University West Lafayette, Indiana 47907-2054

CORN (ZEA MAYS P0157AM)

A trial was established at the Northeast Purdue Agricultural Center (NEPAC) in Whitley County, Indiana. The experiment was a randomized complete block design with four replications. Plots were 30 feet wide and 360 feet long and consisted of four rows, and the center rows were used for evaluation. The previous crop was soybean. Standard practices for nonirrigated grain corn production in Indiana were followed. Corn hybrid P0157AM was planted in 30-inch row spacing at a rate of 32,000 seeds/acre on May 6. Fungicide treatments were applied on June 25 at V6, July 13 at V10, July 20 at tassel/silk (VT/R1), August 4 at blister (R2), and August 18 at dough (R4) growth stages. Little to no diseases were detected in the trial. The trial was harvested on October 7, and yields were adjusted to 15.5% moisture. Data were subjected to mixed model analysis of variance (SAS 9.4, 2019), and means were compared using Fisher's Least Significant Difference test (LSD; α=0.05).

In 2020, very little disease developed in plots. Gray leaf spot (<10%), tar spot (0.01%), southern rust (0.01%), and northern corn leaf blight (<1%) were noted at R6 but not rated. There was no significant effect of fungicide timing on moisture and yield (Table 48).

TABLE 48. *Effect of Fungicide on Corn Yield*

TREATMENT, RATE/ACRE, AND TIMING[z]	HARVEST MOISTURE %	YIELD[y] BU/ACRE
Nontreated control	16.4	195.2
Headline AMP 1.68 SC 10.0 fl oz V6	16.6	203.7
Headline AMP 1.68 SC 10.0 fl oz V10	16.5	198.5
Headline AMP 1.68 SC 10.0 fl oz VT/R1	16.6	200.1
Headline AMP 1.68 SC 10.0 fl oz R2	16.3	196.1
Headline AMP 1.68 SC 10.0 fl oz R4	16.4	195.9
P-value[x]	*0.8887*	*0.7188*

[z] Fungicide treatments were applied on June 25 at V6, July 13 at V10, July 20 at tassel/silk (VT/R1), August 4 at blister (R2), and August 18 at dough (R4) growth stages.

[y] Yields were adjusted to 15.5% moisture at harvest on November 7.

[x] Means followed by the same letter are not significantly different based on Fisher's Least Significant Difference test (LSD; α=0.05).

FIELD-SCALE FUNGICIDE TIMING COMPARISON FOR FOLIAR DISEASES ON SOYBEAN IN NORTHEASTERN INDIANA, 2020 (SOY20-12.NEPAC)

D. E. P. Telenko, J. D. Ravellette, and S. Shim, Department of Botany and Plant Pathology, Purdue University West Lafayette, Indiana 47907-2054

SOYBEAN (*GLYCINE MAX* P35T75X)

A trial was established at the Northeast Purdue Agricultural Center (NEPAC) in Whitley County, Indiana. The experiment was a randomized complete block design with four replications. Plots were 30 feet wide and 400 feet long. The previous crop was corn. Standard practices for nonirrigated soybean production in Indiana were followed. Soybean cultivar P35T75X was drilled in 7.5-inch row spacing at a rate of 150,000 seeds/acre on May 8. Fungicides were applied on June 29 at the fourth-leaf (V4), July 13 at the beginning flower (R1), August 4 at the beginning pod (R3), and August 18 at the beginning seed (R5) growth stages. Little to no disease developed in the field. The soybeans were harvested on October 16, and yields were adjusted to 13% moisture. Data were subjected to mixed model analysis of variance (SAS 9.4, 2019), and means were compared using Fisher's Least Significant Difference test (LSD; α=0.05).

In 2020, very little disease developed in plots. There was no significant effect of treatment on soybean yield (Table 49).

TABLE 49. *Effect of Fungicide on Soybean Yield*

TREATMENT, RATE/ACRE, AND TIMING[z]	HARVEST MOISTURE %	YIELD[y] BU/ACRE
Nontreated control	11.0 b	75.2
Miravis Top 1.67 SC 13.7 fl oz at V4	11.2 b	75.3
Miravis Top 1.67 SC 13.7 fl oz at R1	11.5 a	76.5
Miravis Top 1.67 SC 13.7 fl oz at R3	11.1 b	76.1
Miravis Top 1.67 SC 13.7 fl oz at R5	11.2 a	78.0
P-value[x]	0.0222	0.7681

[z] Fungicide treatments were applied July 10 at the fourth-leaf (V4), July 13 at the beginning flower (R1), August 6 at the beginning pod (R3), and August 29 at the beginning seed (R5) growth stages. All treatments contained a nonionic surfactant (Preference) at a rate of 0.25% v/v.

[y] Yields were adjusted to 13% moisture at harvest on October 16.

[x] Means followed by the same letter are not significantly different based on Fisher's Least Significant Difference test (LSD; α=0.05).

SOUTHEAST PURDUE AGRICULTURAL CENTER (SEPAC)

FIELD-SCALE FUNGICIDE TIMING COMPARISON FOR FOLIAR DISEASES ON CORN IN SOUTHEASTERN INDIANA, 2020 (COR20-11.SEPAC)

S. Shim, J. D. Ravellette, and D. E. P. Telenko, Department of Botany and Plant Pathology, Purdue University West Lafayette, Indiana 47907-2054

CORN (*ZEA MAYS* P0574AM)

Southern rust, *Puccinia polysora*
Gray leaf spot, *Cercospora zeae-maydis*

A trial was established at the Southeast Purdue Agricultural Center (SEPAC) in Jennings County, Indiana. The experiment was a randomized complete block design with four replications. Plots were 30 feet wide and 550 feet long and consisted of 12 rows, and the two center rows were used for evaluation. The previous crop was soybean. Standard practices for nonirrigated corn production in Indiana were followed. Corn cultivar P0574AM was planted in 12-inch row spacing at a rate of 30,000 seeds/acre on May 13. All fungicide applications were applied at 20 gal/acre and 40 psi using an Apache 720 sprayer. Fungicides were applied on June 26 at the V6, July 8 at the V10, and July 28 at the late silking (R1) growth stages. Disease ratings were assessed on September 7 at the maturity (R6) growth stage. Southern rust (SR) and gray leaf spot (GLS) were rated for disease severity by visually assessing the percentage of symptomatic leaf area on the ear leaf on September 7. Yields were adjusted to 15.5 % moisture. Data were subjected to mixed model analysis of variance (SAS 9.4, 2019), and means were compared using Fisher's Least Significant Difference test (LSD; α=0.05).

In 2020, very little disease developed in plots. Southern rust (SR) and gray leaf spot (GLS) were the most prominent diseases but reached low severity. There were no significant differences between fungicide application timing and nontreated control for SR and GLS on September 7 (Table 50). There was no significant treatment effect on corn yield.

TABLE 50. *Effect of Fungicide Timing on Foliar Diseases and Corn Yield*

TREATMENT AND RATE/ACRE[z]	SR % SEVERITY[y]	GLS % SEVERITY[y]	HARVEST MOISTURE %	YIELD[x] BU/ ACRE
Nontreated control	3.8	4.8	24.9 b	210.3
Lucento 4.17 SC 5.0 fl oz at V6	5.6	7.1	25.7 a	210.4
Lucento 4.17 SC 5.0 fl oz at V10	3.9	4.2	25.5 a	212.7
Lucento 4.17 SC 5.0 fl oz at R1	2.8	4.1	25.5 a	210.4
P-value[w]	0.4537	0.6492	0.0085	0.9032

[z] Fungicide treatments were applied on June 26 at the V6, July 8 at the V10, and July 28 at the late silking (R1) growth stages. All treatments contained a nonionic surfactant (Preference) at a rate of 0.25% v/v.

[y] Disease severity was visually assessed as a percentage (0–100%) of symptomatic leaf area on ear leaf on September 7. SR = southern rust, GLS = gray leaf spot.

[x] Yields were adjusted to 15.5 % moisture at harvest on September 21.

[w] Means followed by the same letter are not significantly different based on Fisher's Least Significant Difference test (LSD; α=0.05).

FIELD-SCALE FUNGICIDE TIMING COMPARISON FOR FOLIAR DISEASES ON SOYBEAN IN SOUTHEASTERN INDIANA, 2020 (SOY20-11.NEPAC)

C. Rocca Da Silva, S. Shim, J. D. Ravellette, and D. E. P. Telenko, Department of Botany and Plant Pathology, Purdue University West Lafayette, Indiana 47907-2054

SOYBEAN (*GLYCINE MAX* P35T75X)

Frogeye leaf spot, *Cercospora sojina*
Cercospora leaf blight, *Cercospora kikuchii/C. flagellaris*

A trial was established at the Southeast Purdue Agricultural Center (SEPAC) in Jennings County, Indiana. The experiment was a randomized complete block design with four replications. Plots were 30 feet wide and 700 feet long and consisted of 12 rows, and the two center rows were used for evaluation. The previous crop was soybean. Standard practices for nonirrigated soybean production in Indiana were followed. Soybean cultivar P34A79X was planted in 12-inch row spacing at a rate of 30,000 seeds/acre on June 8. All fungicide applications were applied at 20 gal/acre and 40 psi using an Apache 720 sprayer. Fungicides were applied on June 8 at the V4, August 7 at the beginning pod (R3), and August 26 at the beginning seed (R5) growth stages. Disease ratings were assessed on September 7 at the full seed (R6) growth stage. Frogeye leaf spot (FLS) and Cercospora leaf blight (CLB) were rated for disease severity by visually assessing the percentage of symptomatic leaf area in the upper canopy on September 7. Soybeans were harvested on October 6, and yields were adjusted to 13% moisture. Data were subjected to mixed model analysis of variance (SAS 9.4, 2019), and means were compared using Fisher's Least Significant Difference test (LSD; α=0.05).

In 2020, very little disease developed in plots. FLS and CLB were the most prominent diseases but only reached low severity. There was no significant treatment differences on severity of FLS and CLB (Table 51). Lucento applied at R3 and R5 had significantly higher soybean yield over the nontreated control and Lucento applied at V4.

TABLE 51. *Effect of Fungicide on Foliar Diseases and Soybean Yield*

TREATMENT, RATE/ACRE, AND TIMING[z]	FLS % SEVERITY[y]	CLB % SEVERITY[y]	YIELD[x] BU/ACRE
Nontreated control	0.4	0.1	70.0 b
Lucento 4.17 SC 5.0 fl oz at V4	0.3	0.1	70.0 b
Lucento 4.17 SC 5.0 fl oz at R3	0.5	0.3	79.7 a
Lucento 4.17 SC 5.0 fl oz at R5	0.0	0.0	76.1 a
P-value[w]	0.2797	0.7375	0.0002

[z] Fungicide treatments were applied on June 8 at the V4, August 7 at the beginning pod (R3), and August 26 at the beginning seed (R5) growth stages. All treatments contained a nonionic surfactant (Preference) at a rate of 0.25% v/v.

[y] Foliar disease severity was visually rated on a scale of 0–100% of canopy with disease symptoms on September 7. FLS = Frogeye leaf spot, CLB = Cercospora leaf blight.

[x] Yields were adjusted to 13% moisture at harvest on October 6.

[w] Means followed by the same letter are not significantly different based on Fisher's Least Significant Difference test (LSD; α=0.05).

APPENDIX: WEATHER DATA

TABLE 1. Average Monthly Conditions at the Purdue Agronomy Center for Research and Education (ACRE), the Pinney Purdue Agricultural Center (PPAC), and the Southwest Purdue Agricultural Center (SWPAC) in Indiana, 2020

MONTH	ACRE			PPAC			SWPAC		
	TEMP. MIN.[x] °F	TEMP. MAX.[x] °F	TOTAL PRECIPIT.[y] (IN)	TEMP. MIN.[x] °F	TEMP. MAX.[x] °F	TOTAL PRECIPIT.[y] (IN)	TEMP. MIN.[x] °F	TEMP. MAX.[x] °F	TOTAL PRECIPIT.[y] (IN)
January	18.73	34.85	5.60	16.39	31.74	5.60	24.34	41.01	5.51
February	19.48	37.03	5.37	17.25	33.77	5.37	25.57	43.34	5.44
March	31.45	51.07	7.89	28.21	46.90	7.89	36.50	55.99	7.70
April	39.98	63.28	10.43	37.15	59.77	10.43	46.42	67.78	10.18
May	51.86	74.49	11.44	48.63	70.59	11.44	56.53	77.26	10.99
June	60.98	82.76	10.73	58.68	80.32	10.73	65.01	86.50	10.14
July	62.80	84.47	9.69	60.97	82.78	10.15	67.43	88.40	10.15
August	60.71	83.51	9.46	59.04	81.29	9.46	65.92	88.22	9.76
September	54.24	79.24	7.91	52.68	76.83	7.91	59.22	83.26	8.01
October	43.14	65.87	8.59	41.67	63.12	8.59	47.85	69.99	8.59
November	34.08	52.83	8.51	32.31	49.73	8.51	37.62	56.91	8.51
December	24.56	39.57	7.40	22.47	36.59	7.40	29.14	44.82	7.18

[z] Data courtesy of the Indiana State Climate Office, Beth Hall and Jonathan Weaver, https://ag.purdue.edu/indiana-state-climate/ Taken from the Purdue Mesonet stations at the Purdue Agronomy Center for Research and Education (ACRE), the Pinney Purdue Agricultural Center (PPAC) and the Southwest Purdue Agricultural Center (SWPAC).

[y] Average minimum and maximum temperatures for each month.

[x] Total precipitation for each month.

ABOUT THE AUTHOR

DARCY E. P. TELENKO is an associate professor and Extension plant pathologist in the Department of Botany and Plant Pathology at Purdue University. Her interdisciplinary research and Extension program are involved in studying the biology and management of soilborne and foliar pathogens of agronomic crops. Telenko is a native of western New York and received her PhD at North Carolina State University. She has published more than sixty peer-reviewed manuscripts and two hundred Extension publications. She was awarded the 2024 Leadership Award from the Purdue University Cooperative Extension Specialist Association.

www.ingramcontent.com/pod-product-compliance
Lightning Source LLC
Chambersburg PA
CBHW081139170526
45165CB00008B/2728